PRAISE FOR *THE STEM SHIFT* BY ANN MYERS AND JILL BERKOWICZ

"*The STEM Shift* provides an excellent overview for a school or district contemplating a move to a STEM learning approach. The authors make a strong case for engaging in a STEM shift and then step through the planning and change processes called for, with special attention to building collaboration and trust. A real strength of the book is the inclusion of real cases and voices from the field where this shift is happening. The inclusion of successful programs and perspectives gives the reader confidence that engaging in a STEM shift is a real possibility for their organization."

—Peter Brouwer, Dean, School of Education &
Professional Studies and Graduate Studies

"Myers and Berkowicz expose the power of STEM and how it heightens the trajectory for student success. *The STEM Shift* is the definitive go-to guide for educators."

—Leigh Byron, Founder and
Head of School of STEM-to-Civics Charter School

"*The STEM Shift* frames the current STEM movement as a powerful lever for the learning of all students (and their teachers) and issues a call to leadership to make it so. The book offers both a useful primer on what STEM means and an expansive vision as to what it can be. Both will serve teachers, principals, superintendents, board members and others who will drive this change."

—Terry Chadsey, Executive Director of
the Center for Courage & Renewal

"Want to know the four most important words for 21st century learning? Authors Myers and Berkowicz reveal that *science, technology, engineering* and *math* are the four words that will unlock your students' learning opportunities and open up a world of possibilities. This book is filled with examples that will help schools and districts become 'STEM-centric' learning environments focused on identifying and solving problems. *The STEM Shift* will help you build bridges to the 21st century . . . and beyond!"

—Spike C. Cook,
Author of *Connected Leadership: It's a Click Away*

"*The STEM Shift* is a much needed book for K–20 educational institutions looking to move from good to great. It highlights the importance of why we need intentional learning environments that drive innovation and collaboration."

—Brad Currie, Lead Learner, Co-Founder of
#Satchat, ASCD Emerging Leader, Author of *All Hands on Deck:
Tools for Connecting Educators, Parents, and Communities*

"This insightful book casts light on the important issue of teaching STEM in a coherent and integrated way. The strategies presented go beyond curricula in each discipline to focus on four design principles: subject integration, project-based learning, relevancy, and structural flexibility. This is an important, useful resource for anyone interested in STEM."

—Chris Dede, Wirth Professor in Learning Technologies

"In these days of initiative fatigue, it is easy for leaders and teachers to bypass all that is good about our changing education system. If they do, they are missing great opportunities to advance. Unfortunately, STEM is sometimes seen as the new four-letter word in initiatives and it shouldn't be. In *The STEM Shift*, Ann Myers and Jill Berkowicz offer compelling reasons why all schools need to make the shift and provide practical tools that will help schools get there."

—Peter DeWitt, Corwin Consultant and Author of
Flipping Leadership Doesn't Mean Reinventing the Wheel

"The authors provide examples from real leaders across the country who have transformed their school to increase STEM instruction. This book will make readers think about ways to revolutionize schools by embracing the critical importance of STEM education."

—Lisa Dieker, Professor and Lockheed Martin Eminent Scholar Chair

"Reading Jill and Ann's column in *Education Week* has been a critical part of my weekly reading since they began writing it. I've learned a lot from those short snippets, and now it's exciting to see their expanded thoughts in *The STEM Shift*. You can't go wrong by reading anything they write!"

—Larry Ferlazzo, High School Teacher,
Education Week Columnist, and Author of *Helping Students Motivate Themselves*

"Myers and Berkowicz have done a masterful job of articulating the current and future need for all to engage in conversations about STEM. They outline the importance of using inclusive, collaborative, and innovative engaged learning pedagogies and strategies for all teachers and students. Their final 'call to action' chapter is powerful. I would highly recommend this book to anyone!"

—Susan R. Madsen, Orin R. Woodbury Professor of Leadership & Ethics

"The world is evolving at a feverous pace, and this requires students to think and apply skills differently. Myers and Berkowicz provide a compelling narrative loaded with examples from the field to provide educators and schools with the necessary foundation to provide students with the STEM skills and knowledge they need to succeed in careers while better supporting how they learn best."

—Eric Sheninger, Author of *Digital Leadership* and *UnCommon Learning*

"Finally! A great book that clearly explains what STEM education is, why we need it, and how to do it well. A must-read for all educators, parents, and policymakers."

—Tony Wagner, Author of *The Global Achievement Gap* and *Creating Innovators*

The STEM Shift

A Guide for School Leaders

Ann Myers

Jill Berkowicz

A SAGE Company

FOR INFORMATION:

Corwin
A SAGE Company
2455 Teller Road
Thousand Oaks, California 91320
(800) 233-9936
www.corwin.com

SAGE Publications Ltd.
1 Oliver's Yard
55 City Road
London EC1Y 1SP
United Kingdom

SAGE Publications India Pvt. Ltd.
B 1/I 1 Mohan Cooperative Industrial Area
Mathura Road, New Delhi 110 044
India

SAGE Publications Asia-Pacific Pte. Ltd.
3 Church Street
#10-04 Samsung Hub
Singapore 049483

Printed in the United States of America

ISBN: 978-1-4833-1772-4

This book is printed on acid-free paper.

Executive Editor: Arnis Burvikovs
Senior Associate Editor: Desirée A. Bartlett
Editorial Assistant: Andrew Olson
Production Editor: Melanie Birdsall
Copy Editor: Diane DiMura
Typesetter: C&M Digitals (P) Ltd.
Proofreader: Caryne Brown
Indexer: Molly Hall
Cover Designer: Leonardo March
Marketing Manager: Amanda Boudria

15 16 17 18 19 10 9 8 7 6 5 4 3 2 1

Contents

Digital Resources

Access links and videos at http://bit.ly/TheSTEMShift (please note that this URL is case sensitive).

This content will continue to be updated, so please return for new resources and links!

Note From the Publisher

The authors have provided video and web content throughout the book that is available to you through QR codes. To read a QR code, you must have a smartphone or tablet with a camera. We recommend that you download a QR code reader app that is made specifically for your phone or tablet brand.

QR codes may provide access to videos and/or websites that are not maintained, sponsored, endorsed, or controlled by Corwin. Your use of these third-party websites will be subject to the terms and conditions posted on such websites. Corwin takes no responsibility and assumes no liability for your use of any third-party website. Corwin does not approve, sponsor, endorse, verify, or certify information available at any third-party video or website.

PART I. WHY STEM?

Chapter 1. The Tipping Point

Center for Public Education. (2009). *21st Century Skills*: http://www.centerfor publiceducation.org/Main-Menu/Policies/21st-Century/21st-century-demographics-21st-century-skills-.html

Center for Public Education. (2009). *Defining a 21st Century Education*: http://www.centerforpubliceducation.org/Main-Menu/Policies/21st-Century/Defining-a-21st-Century-Education-Full-report-PDF.pdf

Center for Public Education. (2012). *The United States of Education: The Changing Demographics of the United States and Their Schools*: http://www.centerforpub liceducation.org/You-May-Also-Be-Interested-In-landing-page-level/Organizing-a-School-YMABI/The-United-States-of-education-The-changing-demographics-of-the-United-States-and-their-schools.html

Federal Reserve Bank of Dallas. (2013). *Immigrants in the U.S. Labor Market*: http://www.dallasfed.org/assets/documents/research/papers/2013/wp1306.pdf

Forbes. (n.d.). *Top 10 Companies Hiring Foreign Workers, No. 4 IBM*: http://www.forbes.com/pictures/efei45mdli/no-4-ibm/

Gould, S. J., & Eldredge, N. (2013). *Niles Eldredge–Stephen Jay Gould in the 1960s and 1970s, and the Origin of "Punctuated Equilibria"*: http://yhoo.it/1IEAKTt

Information Technology Industry Council, Partnership for a New American Economy, & U.S. Chamber of Commerce. (2012). *Help Wanted: The Role of Foreign Workers in the Innovation Economy* (Report on foreign workers in STEM): http://www.renewoureconomy.org/sites/all/themes/pnae/stem-report.pdf

Kids Count Data Center (U.S. demographic information on children): http://datacenter.kidscount.org/data/tables/103-child-population-by-race?loc=1&loct=1#detailed/1/any/false/36,868,867,133,38/66,67,68,69,70,71,12,72/423,424

Ryan, C. L., & Siebens, J. (2012). *Educational attainment in the United States: 2009* (Report P20-566): http://www.census.gov/prod/2012pubs/p20-566.pdf

Wulf, W. A., Kouzes, R. T., & Myers, J. T. (1996). *Collaboratories: Doing Science on the Internet*: https://www.cs.virginia.edu/people/faculty/pdfs/Collaboratories.pdf

Chapter 2. The 21st Century Learning Environment

Carnegie Mellon University. *Alice: Teaching Programming Through 3D Animation and Storytelling* (object-oriented, 3D programming environment): http://www.cmu.edu/corporate/news/2007/features/alice.shtml **(QR code on page 21)**

EdSurge News. (2013). *Learn to Code, Code to Learn*: https://www.edsurge.com/n/2013-05-08-learn-to-code-code-to-learn

11-Year-Old Girl Uses Science Project to Create Prosthetic Hands for Children. (2014). (Sierra Petrocelli builds a prosthetic hand): http://kdvr.com/2014/07/15/11-year-old-girl-uses-science-project-to-create-prosthetic-hands-for-children/

McNulty, R. J. (2011). *Best Practices to Next Practices: A New Way of "Doing Business" for School Transformation*: http://teacher.scholastic.com/products/scholastic-achievement-partners/downloads/Best_Practices_To_Next_Practices.pdf

Neal, J. (n.d.). *Edgewalkers*: http://edgewalkers.org

Overmyer, J. *Flipped Learning Network:* http://www.flippedclassroom.com

Partnership for 21st Century Skills (P21). *Framework for 21st Century Learning*: http://www.p21.org/our-work/p21-framework

Scratch (programming for ages 8 to 12): http://scratch.mit.edu **(QR code on page 21)**

ScratchJr (programming for ages 5 to 7): http://www.scratchjr.org

Simon, J. (2014). *E-Nabling Sierra*: http://www.3duniverse.org/2014/05/16/e-nabling-sierra/Teach LivE (the mixed reality tool used to help teachers and leaders improve interactions with students and adults): http://teachlive.org **(QR code on page 22)**

Te@chThought: 21st Century Pedagogy: http://www.teachthought.com/technology/a-diagram-of-21st-century-pedagogy/#respond

Wakin, D. J. (2010): *The Valhalla Machine*: http://www.nytimes.com/2010/09/19/arts/music/19ring.html?pagewanted=all&_r=0

What Is STEM? (video): http://youtu.be/AlPJ48simtE

Chapter 3. Clearing the Path

Alter, C. (2014). *Soon There Will Be Female Scientist LEGOs*: http://time.com/2822921/soon-there-will-be-female-scientist-legos/

Finley, K. (2014). *In a First, Women Outnumber Men in Berkeley Computer Science Course*: http://www.wired.com/2014/02/berkeley-women/

Rhodan, M. (2013). *These Are the 30 People Under 30 Changing the World* (article highlighting Britney Wenger): http://ideas.time.com/2013/12/06/these-are-the-30-people-under-30-changing-the-world/slide/britney-wenger/

STEM Integration in K–12 Education (video): http://youtu.be/AlPJ48simtE **(QR code on page 32)**

ThinkProgress.org. (2014). *Women Are Leaving Science and Engineering Jobs in Droves* (women leaving STEM jobs): http://thinkprogress.org/economy/2014/02/13/3287861/women-leaving-stem-jobs/

Voyer, D., & Voyer, S. D. (2014). *Gender Differences in Scholastic Achievement: A Meta-Analysis*: http://dx.doi.org/10.1037/a0036620

Chapter 4. The Achievement Gap

Brown, T. *Design Thinking: Thoughts by Tim Brown* (Web log): http://designthinking.ideo.com/?p=1165

Institute for Inquiry. (1991). *Constructivist Learning Theory*: http://www.exploratorium.edu/ifi/resources/constructivistlearning.html

Chapter 5. Special Populations

Centers for Disease Control. (2014). *CDC Estimates 1 in 68 Children Has Been Identified With Autism Spectrum Disorder*: http://www.cdc.gov/media/releases/2014/p0327-autism-spectrum-disorder.html

Johnson, L. B. (1964). *91st Annual Message to the Congress on the State of the Union* (January 8, 1964): http://www.presidency.ucsb.edu/ws/?pid=26787

Kids Count Data. (2014). *Child Population by Race*: http://datacenter.kidscount.org/data/tables/103-child-population-by-race?loc=1&loct=1#detailed/1/any/false/36,868,867,133,38/66,67,68,69,70,71,12,72/423,424

Lombardi, J. (2004). *Practical Ways Brain-Based Research Applies to ESL Learners* (ESL learning research): http://iteslj.org/Articles/Lombardi-BrainResearch

National Center for Education Statistics. (n.d.). *Fast Facts: English Language Learners*. U.S. Department of Education, Institute of Education Sciences: https://nces.ed.gov/fastfacts/display.asp?id=96

Short, D. J., & Fitzsimmons, S. (2007). *Double the Work: Challenges and Solutions to Acquiring Language and Academic Literacy for Adolescent English Language Learners*: http://carnegie.org/fileadmin/Media/Publications/PDF/DoubletheWork.pdf

STEM Smartbrief: Raising the Bar: Increasing Achievement for All Students: http://successfulstemeducation.org/resources/raising-bar-increasing-stem-achievement-all-students

U.S. Department of Education. (2007). *History: Twenty-Five Years of Progress in Educating Children With Disabilities Through IDEA*: http://www2.ed.gov/policy/speced/leg/idea/history.html

U.S. Department of Education, Office of Special Education Programs. (n.d.). *Building the Legacy: IDEA 2004*: http://idea.ed.gov/explore/view/p/,root, dynamic,TopicalBrief,23

U.S. Department of Education, Office of Special Education Programs, IDEA: http://idea.ed.gov/explore/home

Wei, X., Yu, J. W., Shattuck, P., McCracken, M., & Blackorby, J. (2013). *Science, Technology, Engineering, and Mathematics (STEM) Participation Among College Students With an Autism Spectrum Disorder*: http://www.ncbi.nlm.nih.gov/pmc/articles/PMC3620841/

PART II. SHIFTING

Chapter 6. The Shift Begins Within a Leader

Moyers, B. (2008). *Democracy in America Is a Series of Narrow Escapes, and We May Be Running Out of Luck*: http://www.alternet.org/story/85521/moyers%3A_%27democracy_in_america_is_a_series_of_narrow_escapes%2C_and_we_may_be_running_out_of_luck%27

Chapter 7. Entering the STEM Shift

Battelle. *STEM Education: Growing Tomorrow's Innovators in Science, Technology, Engineering and Math*: http://www.battelle.org/our-work/stem-education

Teaching Institute for Excellence in STEM (TIES): http://www.tiesteach.org

Chapter 8. Planning the Shift

Arizona STEM Network. (2013). *The STEM Immersion Guide for Schools and Districts:* http://stemguide.sfaz.org/wp-content/uploads/2015/01/SFAz_STEM_ImmersionGuide1214.pdf

Delaney, M. (2014). *7 Guidelines for Building a STEAM Program*: http://www.edtechmagazine.com/k12/article/2014/04/7-guidelines-building-steam-program

East Syracuse Minoa Central School District. *Strategic Plan*: http://www.esmschools.org/district.cfm?subpage=24324 **(QR code on page 75)**

Gollwitzer, P. (1999). Implementation intentions: Strong effects of simple plans. *American Psychologist, 54*, 493–503 (research on preparing for obstacles before they arise): http://www.psychology.nottingham.ac.uk/staff/msh/mh_teaching_site_files/teaching_pdfs/C83SPE_lecture3/Gollwitzer%20(1999).pdf

Goochland County Schools. *G21: A Framework for Developing Twenty-First Century Skills and Deeper Learning Experiences*: http://www.glnd.k12.va.us/index/resource/g21/ **(QR code on page 83)**

Goochland County Schools. *Strategic Plan*: http://www.glnd.k12.va.us/index/schoolboard/plan **(QR code on page 83)**

Granite School District Five-Year Framework: http://www.graniteschools.org/teachinglearning/wp-content/uploads/sites/13/2014/11/GSD-5-Year-Plan.pdf **(QR code on page 78)**

Chapter 9. STEM Curriculum Shifting

Buck Institute for Education (BIE). *Project Search* (for curated project-based learning examples): http://bie.org/project_search/results/search/P450

Eesha Khare: Inventing the One Minute Mobile Phone Charger: https://www.youtube.com/watch?v=kMWWOnnS3ZM **(QR code on page 102)**

Mayo, G. (2014). *Creating Architectural Models of Literary Themes* (reader idea; Montgomery Blair High School English Class): http://learning.blogs.nytimes.com/2014/05/15/reader-idea-creating-architectural-models-of-literary-themes/

Montgomery County Public Schools. *Vision, Mission, and Core Values*: http://www.montgomeryschoolsmd.org/boe/about/mission.aspx

New Milford High School. *Program of Studies 2014–2015 Academic Season*: http://www.newmilfordschools.org/NMHS/media/Course_of_Study_2014-15.pdf

Pericoli, M. (2013). *Writers as Architects*: http://opinionator.blogs.nytimes.com/2013/08/03/writers-as-architects/

President Obama Speaks at the 2014 White House Science Fair (video): http://youtu.be/vke-SE1mqIs **(QR code on page 103)**

Project Lead the Way: https://www.pltw.org

Project Lead the Way. (2014). *The Louis Calder Foundation Puts Support Behind Elementary STEM Education*: https://www.pltw.org/news/items/201405-louis-calder-foundation-puts-support-behind-elementary-stem-education

Stratford High School Profile. http://www.mnps.org/pages/mnps/About_Us/MNPS_Schools/High_Schools/High_School_Profile/Stratford_High_School_Profile **(QR code on page 97)**

Stratford STEM Magnet High School (video with Principal Michael Steele) http://youtu.be/4bWFb9vjEXE **(QR code on page 96)**

The Virginia Initiative for Science Teaching and Achievement (VISTA).*What Is VISTA? A Program Overview* (video): http://youtu.be/5M3n3Vlfyog

Chapter 10. Developing Capacity: STEM-Centric Professional Development

Buck Institute for Education (BIE). *Project Search* (for curated project-based learning examples): http://bie.org/project_search/results/search/P450 **(QR code on page 112)**

Charlotte-Mecklenburg Schools. *Strategic Plan 2018*: http://www.cms.k12.nc.us/mediaroom/strategicplan2018/Documents/Strategic%20Plan%202018%20For%20a%20Better%20Tomorrow%20Fact%20Sheet.pdf **(QR code on page 115)**

DuPont Hadley Middle School: http://duponthadleyms.mnps.org/pages/DuPont_Hadley_Middle_School

Middle Tennessee STEM Hub: http://midtnstem.com

National Education Association. *The 10 Best STEM Resources Science, Technology, Engineering & Mathematics Resources for PreK–12*: http://www.nea.org/tools/lessons/stem-resources.html

Project LIFT. *About Project L.I.F.T.* (public/private partnership nonprofit organization, operating as one of five learning communities in the Charlotte-Mecklenburg School System): http://www.projectliftcharlotte.org/about

Public Impact (Helping education leaders and policymakersimprove student learning in K–12 education). *About Public Impact*: http://publicimpact.com/about-public-impact/

Tennessee Department of Education. (2014). *First to the Top*: http://www.tn.gov/education/about/fttt.shtml

The Tennessee Higher Education Commission and Battelle Memorial Institute. *Battelle Memorial Institute* (video): http://youtu.be/8PckHDH_6Ho **(QR code on page 111)**

The Virginia Initiative for Science Teaching and Achievement (VISTA): http://vista.gmu.edu **(QR code on page 116)**

The Virginia Initiative for Science Teaching and Achievement (VISTA). *VISTA Voices* (videos): https://www.youtube.com/user/VISTAScience?feature=watch **(QR code on page 116)**

The Virginia Initiative for Science Teaching and Achievement (VISTA). *What Is VISTA? A Program Overview*: http://youtu.be/5M3n3Vlfyog **(QR code on page 116)**

Chapter 12. STEM Collaborations and Trust

Center for Courage & Renewal. *Circle of Trust Touchstones*: http://www.couragerenewal.org/wpccr/wp-content/uploads/touchstones-poster.pdf **(QR code on page 136)**

Chapter 13. A Call to Action

Columbia University School of the Arts. *The Laboratory of Literary Architecture: A Workshop With Matteo Pericoli* (literary architecture lab at Columbia University): http://arts.columbia.edu/laboratory-literary-architecture-workshop-matteo-pericoli

EdSurge (for science and technology news and updates): https://www.edsurge.com

Edutopia (sharing evidence- and practitioner-based learning strategies that empower you to improve K–12 education): http://www.edutopia.org

edWeb: A professional online community for educators: http://home.edweb.net

Freeman, S., Eddy, S. L., McDonough, M., Smith, M. K., Okoroafor, N., Jordt, H., & Wenderoth, M. P. (2014). Active learning increases student performance in science, engineering, and mathematics. *Proceedings of the National Academy of Sciences of the United States of America, 111*, 8410–8415 (meta-analysis about effect of lecture on learning): http://iteachem.net/wp-content/uploads/2014/05/Freeman-S-Proc-Natl-Acad-Sci-USA-2014-Active-learning-increases-student-performance-in-science-engineering-and-mathematics.pdf

Mayo, G. (2014). *Creating Architectural Models of Literary Themes* (reader idea; Montgomery Blair High School English Class): http://learning.blogs.nytimes.com/2014/05/15/reader-idea-creating-architectural-models-of-literary-themes/

Mind/Shift (launched in 2010 by KQED and NPR; explores the future of learning in all its dimensions): http://blogs.kqed.org/mindshift/

The National Education Association. *The 10 Best STEM Resources: Science, Technology, Engineering & Mathematics Resources for PreK–12*: STEM lesson resources: http://www.nea.org/tools/lessons/stem-resources.html

STEMconnector (information about STEM): https://www.stemconnector.org

Teaching Institute Excellence in STEM (TIES): http://www.tiesteach.org

Preface

STEM has an undeniable presence on the educational landscape. STEM-identified classes, programs, and events are appearing everywhere. States have funded initiatives to develop a generation of STEM teachers. An extraordinary coalition among business leaders, government leaders, philanthropists, parents, and communities at large is generating a crescendo of support for STEM in schools.

STEM seedlings emerged in charter schools, in magnet schools, or as specialized high schools. Supported by government and private dollars, they began to attract attention. Forward-thinking school leaders and their teams traveled to visit them and considered how to replicate them. STEM programs are being created daily across the nation, but few districts are exploring the full depth and breadth of possibility that this book presents.

● DEFINING THE STEM SHIFT

There is a four-word definition for STEM. It involves the four subject areas identified in the acronym . . . science, technology, engineering, and math. The name and the initiative emanate from the National Science Foundation over fifteen years ago, but the seeds for the STEM shift were planted in the 1950s after *Sputnik*. Since then, STEM has grown slowly and organically without sensationalism or the power of legislation and regulation. Now, it is sprouting everywhere.

Our definition of STEM is inspired by the teachers and leaders of the Metro Nashville School District. We see **STEM as a shift in the philosophical framework for teaching and learning.** The shift leaves behind a subject-based, rigidly scheduled, unintegrated system to become one that is **defined by subject integration, project-based learning, relevancy for the lives of children, and structural flexibility.**

STEM-based subject integration challenges the current nature of our "siloed" educational systems and encourages teachers to work together in inter- and trans-disciplinary ways. New interactions emerge causing changes in relationships among subject areas, teachers, students, and communities. From the interests and the interaction of teachers and students, problems and projects are created. Project-based learning involves students in teams actively working toward solving complex, real-world problems. Relevancy relates content and projects to the lives of our students. Classroom doors open to the community, drawing in STEM professionals as guests, coteachers, and assessment designers and participants. Our vision of STEM brings students into the community (and beyond) through technology to gain experiences and expertise heretofore unavailable. This new system is characterized by organizational and structural flexibility. Schedules no longer constrain the thinking of educators and learners as we become designers of new environments where real-world problems are the focus of student learning

This book investigates STEM as a systemic shift, beginning with revisiting how learning happens and culminating in new learning environments, new structures, and new relationships. The shift relies on access to technology at all levels. It offers the opportunity to reignite the creativity of school leaders and teachers. The interplay of these dynamics, with planning, leadership and resources, becomes a systemwide STEM shift.

As we traveled and listened to the stories of the STEM pioneers, we heard a theme of emergence from within each system. The STEM shift may have been externally set in motion but it was being defined internally. We developed the following chart (see Figure P.1) to describe the emergence of a STEM shift on three levels. We call them Stages I, II, and III. The column on the left identifies who is involved with the interactions shifting at each stage. Moving from top to bottom, more of the school and community are involved. The column on the right describes the characteristics emerging in that stage. The characteristics column is cumulative from top to bottom; therefore, a Stage II shift includes the characteristics of Stage I. Within a school or district in a Stage III shift, all characteristics in the right column will be evident.

You will note that the chart allows entry at any of the three stages. Stage I describes an individually led program that affects a class of students. Stage II is defined by the inclusion of several classrooms, multiple teachers, and STEM field professionals. The schoolwide system STEM shift is Stage III. Remember a district can choose the system level entry stage. Three of the districts described in this book did that. One of the differences between a STEM shift and other "reform efforts" is that each teacher, school, and district described in this book decided how fully they wanted to enter the shifting process. Many are following a plan to move from one to another.

Figure P.1   The Emergence of a STEM Shift

Stages	Characteristics Emerging
I Stand-Alone • Individually led • Leader supported • Small number of students involved	Science, technology, engineering, math (and art) centric 21st century skills Project-based learning Student-directed discovery Teamwork among students Unit-specific community partners Events highlight student work
	+
II Collaborative and Cross Disciplinary • Multiple teachers participate in planning and delivery • Students from multiple classes or classrooms involved	Interdisciplinary or transdisciplinary content STEM professionals as partners in classrooms, labs, events Students and worldwide community engaged digitally STEM experts participate with teaching and learning Student presentation of culminating work
	+
III Systemwide • Planned buildingwide or districtwide STEM shift • All teachers • All students	Culture of innovation, risk taking, and reflection Complex, real-world problem solving Crosses grade levels vertically and horizontally Authentic assessments aligned with STEM Collaborative faculty learning, lesson design, and teaching Ongoing, purposeful professional development Structural flexibility supports opportunity generation Community-, business-, collegiate-embedded partnerships Wholehearted leaders who see the STEM shift as opening possibilities for all children and for the future. They ignite followership, develop capacities, and create energy to sustain the shifting process.

Enter Here or → (to Stage I)

Enter Here or → (to Stage II)

Enter Here → (to Stage III)

● THE AUDIENCE FOR THIS BOOK

This book is written for those who want to make informed choices about STEM. Leaders who want to consider making a schoolwide or districtwide STEM shift will find stories from those who have successfully achieved this. Teachers and teacher leaders are offered experiences of colleagues at elementary, middle school, and high school levels who have been pioneers of the STEM shift. Policymakers and school board members will be able to explore implications for fiscal, physical, and human resource issues as the shift begins and disrupts existing policies and procedures. Parents who chose to become STEM advocates will find support and resources. All will become better prepared as ambassadors for STEM-centric learning environments.

● HOW THIS BOOK IS STRUCTURED

The book is structured for a diverse audience. Part I, Why STEM? contextualizes the need for STEM in the larger world and within the world of education. It addresses the limiting mental models that may be encountered by those wanting to lead a STEM initiative. Part I introduces STEM as both responsive and future defining, as high rigor, high relevance, and inclusive, as a national imperative, and as locally created and controlled.

Part II, Shifting, invites the reader into the process with practical steps, processes, and stories from those who are already shifting. This part begins with a chapter on leadership. It is purposefully placed centrally within the book. Each story we heard wove a tale of leadership. In some cases, it was an entrepreneurial teacher, in others, it was a visionary district leader but, in all, leaders were essential. We discovered edge walkers, those who have ventured out of the familiar terrain to discover potential in unexplored territory. Choices are locally made, assessment is inherent in the design, and students and communities are the beneficiaries.

The stories come from across the country . . . from Nashville, Tennessee; Salt Lake City, Utah; East Syracuse, New York; Goochland County, Virginia, and elsewhere; from elementary, middle, and high schools, and from public schools in a variety of forms: community schools, magnet schools, and vocational schools. As you read them, you will hear how STEM characteristics within a single classroom vary greatly from those of a whole school or district. Referencing the chart above will help you determine where the storyteller is located . . . and maybe where you are or want to be . . . in the shift process. Some are high school programs that focus on the careers in STEM fields. Others have comprehensively shifted systems, beginning in kindergarten and going through high school.

As career educators, our own entry to the STEM world has been exciting. We have been inspired by those who are the pioneers, and we have become increasingly optimistic about education's capacity to reinvent itself, one school at a time. STEM shifts can reenergize professional educators, can engage children as problem solvers, prepare them to be college and career ready, tap the community as resources, and discover corporations as partners. All the potential for a STEM shift is in your hands.

● HIGHLIGHTS

This book contextualizes and defines the current educational landscape, takes a look at the horizon, delineates how to lead a STEM shift, and shares rich stories from those around the nation who have entered, planned, and implemented a STEM shift. Readers will discover

- The converging forces that cause STEM to be at forefront of the educational conversation, with support rather than deep controversy

- An understanding of what STEM means and how it can shift the learning environments into the 21st century for all students

- The importance of a STEM shift in closing the achievement gap

- The essential elements of leadership and leaders

- Their own readiness and their local questions

- Themselves actively engaged in the reconsideration of how they teach and how children learn

- References and Resources at end of each chapter

- QR codes for some resources that can be found within the text

 - Live links and URLs mentioned throughout the book that can be also be accessed by visiting **http://bit.ly/TheSTEMShift** (please note that this URL is case sensitive).

Acknowledgments

We begin with the acknowledgment of those pioneers leading a STEM shift in communities across the country. Especially, we would like to thank the faculty of Hattie Cotton Elementary School and the Metro Nashville K–12 STEM Instructional Design team. They shared the successes, trials, hopes, and losses of a two-year professional and personal journey as they lived through a STEM shift. We are respectful and indebted. Vicki Metzgar, Executive Director, Middle Tennessee STEM Innovation Hub, served as a valuable resource during the writing of this book. Her love of science, vision for a STEM shift, unrelenting effort to make the shift come alive for Metro Nashville students, personal authenticity, and professional leadership was inspirational to us.

Specifically, we acknowledge those who took time from their work as teachers and leaders to share their experiences with us. They came from Goochland County Public Schools in Virginia, Charlotte-Mecklenburg Schools in North Carolina, Granite School District in Utah, Boston Green Charter School in Massachusetts, New Milford School District in New Jersey, Metro Nashville Public Schools in Tennessee, Montgomery County Public Schools in Maryland, SUNY New Paltz in New York, East Syracuse Minoa Central Schools in New York, and the ORT Israel Network in Israel. We also thank Jeff Doran, PhD, who listened to our conversations about the STEM shift and recognized in it the parallel to punctuated equilibrium in his world of science.

A special thank you goes to Arnis Burvikovs for seeing the value in developing our ideas into this book with Corwin, to Desirée Bartlett, Ariel Price, and Melanie Birdsall for their help and support throughout the process and to Andrew Olson, who helped attend to the details of the process. Thank you, also, to Pat Loehfelm, who accurately, and with interest, transcribed all of the interviews. During the writing of this book, we found our way to the Carondelet Hospitality Center and the St. Joseph's Center, where we were welcomed by sisters and brothers, into generative writing spaces, with meals to sustain our energy.

And finally, we extend our deepest gratitude to Ed Hallenbeck and Bob Santoro. They were unwavering in their support of us in this project. They broke our intensity with laughter and gifted us with patience. These kind, honest, gentle men live their lives in service to others and give of themselves joyfully. We benefit from them and count them as blessings in our lives.

● PUBLISHER'S ACKNOWLEDGMENTS

Corwin gratefully acknowledges the contributions of the following reviewers:

Eric T. Armbruster
Elementary Principal
Henrico County Public Schools
Henrico, VA

Laurie Barron
Superintendent
Evergreen School District
Kalispell, MT

Peter Brouwer
Dean
SUNY Potsdam
    School of Education
Potsdam, NY

Peter DeWitt
Education Consultant
    and Corwin Author
Albany, NY

Jeanne Gren
Principal
Flemington Elementary
Flemington, WV

Mary Ann Hartwick
Principal
Litchfield Elementary School
Litchfield Park, AZ

Freda Hicks
Principal
Perry Harrison School
Pittsboro, NC

Martin Hudacs
Educational Consultant
Quarryville, PA

Barbara Malkas
Superintendent
Webster Public Schools
Webster, MA

About the Authors

Ann Myers, EdD, is Professor Emerita at Sage Colleges, where she was Founding Director of the Sage Doctoral Program in Educational Leadership and of the Dawn Lafferty Hochsprung Center of the Promotion of Mental Health and School Safety. Prior to her career in higher education, she was District Superintendent at Questar III BOCES, a regional educational organization in Upstate New York. She is a national facilitator affiliated with the Center for Courage and Renewal. She leads Circle of Trust® retreats, leadership development programs, and strategic planning for school districts and nonprofit organizations. Ann was a consultant for the Metro Nashville Public Schools on the trust building project. She is the founding chair of New York State Association for Women in Administration and remains on the board for that organization. To all of this work, Ann brings a unique mix of intelligence, wisdom, and intuition. She brought an understanding of the nature of an edge walker, called to innovation and exploration and to reflection and growth for self and others. Ann is coauthor with Jill Berkowicz, of *Leadership 360*, a blog published by Education Week. Ann lives between the Adirondacks in New York State and the Gulf Coast of Florida with her husband Ed Hallenbeck and her golden retriever.

Jill Berkowicz, EdD, has spent her thirty-year career in education focusing on issues of equity and best practices in curriculum, instruction, assessment, and technology, as they affect the development of all learners in the K–12 system. In her leadership roles as a principal and director of curriculum, instruction, and technology, equal access to quality teaching was the foundation of her work. Presently, Jill is an adjunct professor at SUNY New Paltz in the educational leadership program. She provides ongoing professional development for teachers and principals. Her work is dedicated to developing schools' capacity to improve the teaching and learning environment through technology, high-stakes port-folios, and rigorous learning experiences. She serves on the board of New York ASCD and as the educational consultant for their digital newsletter. Jill brings energy, passion for innovation and collaboration to all her work. Jill is coauthor, with Ann Myers, of *Leadership 360*, a blog published by Education Week. Jill lives in the Hudson Valley of New York with her partner Bob Santoro.

Part I

Why STEM?

Here's to the crazy ones. The misfits. The rebels. The troublemakers. The round pegs in the square holes. The ones who see things differently. They're not fond of rules. And they have no respect for the status quo. You can quote them, disagree with them, glorify or vilify them. About the only thing you can't do is ignore them. Because they change things. They push the human race forward. And while some may see them as the crazy ones, we see genius. Because the people who are crazy enough to think they can change the world, are the ones who do.

—Steve Jobs (Apple Inc.)

1 The Tipping Point

The tipping point is that magic moment when an idea, trend, or social behavior crosses a threshold, tips, and spreads like wildfire.

—Malcolm Gladwell

The past decades have presented multiple attempts at education reform. Still, the demands on the system for change endure. Those demands come from multiple sources and are multifaceted. They arrive with unrelenting speed. Think time has vanished to the point where it seems a luxury for leaders. As a result, wise, committed, innovative educational leaders have become master tinkerers. Neither the demands nor our responses have resulted in a reconceptualization of public education.

Past reform efforts led to relatively minor increases in performance. It is not surprising that the role of the federal government in education became more heavy-handed as the 20th century turned a corner. United States students were slipping in performance when compared to students from the Far East and Finland. Bipartisan support led to the passage of the No Child Left Behind (NCLB) act in 2001. The shortcomings of public education, the need for increased accountability, and a zoom lens on teachers became rallying calls for reform mandates and political agendas nationally.

The debates, like many others currently ongoing in our country, polarized the stakeholders. Common Core Standards, related assessments and teacher and principal performance evaluations, global competition, charter school competition, and declining resources made for a perfect storm.

● DATA AND LEGISLATION

Educators knew there was a soft underbelly about which they did not speak. The achievement gap appeared on center stage as agencies began reporting the data by subgroup population performance. The focus on high school graduation rates and college entry now was accompanied by a conscience factor. Children in poverty were receiving a second-class education. According to the 2013 Kids Count data, that population comprises 40 percent of Black and African American children, 37 percent of American Indian children and 34 percent of Hispanic and Latino children and compared to 14 percent of non-Hispanic white children (see Kids Count). The school-age population is increasingly bilingual as well. Immigration and birth rates combine to accelerate the nation's schools to the shifting point where there will no longer be a non-Hispanic white majority. It is these very children with whom schools have been least successful. Alarms went off. The goal of No Child Left Behind was to eliminate the achievement gap associated with race and social class (Rothstein, 2004). NCLB set a goal that proficiency for all was to be reached by 2013–2014. That date has come and gone. The achievement gap remains.

The U.S. Census Bureau's report, *Educational Attainment in the United States: 2009*, indicates that only 60.9 percent of Hispanic students complete high school or more, for example, compared to 90.4 percent of non-Hispanic whites and 81.4 percent of African Americans (see Ryan & Siebens, 2012). The most difficult and legitimate complaint against our system emerged from this kind of data.

Fiscal policy exacerbates the problem. State funding does not flow based on the needs of the children served. The Great Recession of 2008 resulted in cuts to state funding for education across the country. Tax caps were passed to alleviate tax burdens, as school funding and public employee benefits became hot spots in the political arena.

Race to the Top was passed in 2009. Its purposes were to increase all student achievement, eliminate existing achievement gaps, increase the graduation rates, and produce graduates who were college and career ready to compete in the global economy. States desiring access to the funding associated with the law adopted the Common Core Standards, created new testing and tracking systems, and incorporated student results into personnel evaluation systems.

That final provision is most revealing. It touched the heart of every school with teacher and principal evaluation. Now, the federal government was not just lifting up expectations for students; it was also threatening the work security of those who were not making enough contribution to that effort. Funds were included to entice states and districts to "choose" compliance. In desperate fiscal conditions, due to the loss of state revenues and the repercussions at the local level, most did.

Educators, parents, and policymakers alike knew for years that not all children were receiving equal educations. But it was politically expedient and purposefully essential to maintain public support. America's schools are, at least most of them, publicly funded and responsible to children, parents, and taxpayers. Therefore, educational leaders became masters at discussing and publishing those data that were positive.

The percentage of graduates going on to higher education was one of those. The open enrollment admission policies of many colleges bolstered the numbers. Then, those very same institutions began complaining about the preparation of their freshmen. All the positive data was true, but it was also incomplete. As others began to reveal more negative data, efforts to retain public support made educators look as if they were purposely hiding something. All were hampered by the complexity of the issue and the dimensions of the problem. Even with no personal gain or systemic benefit at stake, trust for the system and its leaders eroded.

● SOCIAL AND ECONOMIC FACTORS

In addition to the purely educational issues, there are social and economic issues driving the agenda. The nation's population is aging rapidly. The United Nations Population Division coined the phrase *gray tsunami* to describe the growing percentage of the population that is over sixty years of age and living well into their eighties. These Americans will be dependent on the growing immigrant population in the United States, as they become the workforce on which our economy, and our democracy, will rely. Of course, minority students' lower achievement in science and math presents a jeopardized feeder system for professions in those fields.

The Federal Reserve Bank of Dallas (2013) reports the immigrant population had increased from twenty-five million in 1996 to forty million by 2011. So it is not surprising that the Pew Research Center notes that an astounding 93 percent of the workforce expansion over the next thirty-six years will be made up of new immigrants and their second-generation offspring. As immigrant population numbers increase, the United States school population reflects that changing landscape. According to the Center for Public Education (2012),

The U.S. Census Bureau predicts that "minorities" will make up the majority of U.S. schoolchildren by 2023, the majority of working-age Americans by 2039, and the majority of all Americans by 2042 . . . American students can expect to live and work in communities that will be much more diverse.

Within a few decades, the school-age population will be primarily black, Hispanic, and Asian; the elder population will be primarily non-Hispanic white. The profile of race in America will be profoundly evident when the age factor is considered.

● WORKPLACE AND TECHNOLOGY

The need for a highly skilled workforce and the lack of success with the very populations that will constitute the workforce of the future captured the attention and heightened the concerns of business leaders. They are already seeking foreign workers in STEM professions at increasing rates. H-1B visas allow companies to employ foreign workers in occupations that require highly specialized knowledge in fields such as science, engineering, computer programming, medicine, health, and economics, among others. Those workers can be employed in the United States for full-time jobs for up to six years. Companies like Microsoft, Facebook, and Google have advocated increasing the number of those visas allowed each year, indicating they cannot find American workers for these jobs. In fact, Facebook is called a **"visa dependent" company** with 15 percent of its workers from other countries. But they are not alone.

Forbes reports that IBM hired 6,190 highly skilled immigrant workers in 2012 and paid them an average annual salary of $82,630 (see Forbes). In the same year, Microsoft employed over 4,000 foreign workers. They received an average annual salary of over $109,000. From big to small, businesses are clamoring for a more highly skilled workforce prepared to enter STEM professions. A 2012 report from Information Technology Industry Council, Partnership for a New American Economy, and the U.S. Chamber of Commerce suggests that "by 2018 there will be more than 23,000 advanced degree STEM jobs that will not be filled even if every new American STEM grad finds a job" (p. 1). In addition, this report discloses that currently there is full employment for U.S. workers with advanced STEM degrees and that STEM fields hire a higher proportion of foreign workers than non-STEM fields.

Simultaneously, the cost of a computer has decreased as computer capacity has increased. More people own laptops, and mobility has

> STEM possesses the first really promising potential to reenvision the educational orientation from the bottom up.

become an asset. More businesses offer wireless service to their customers. *Wi-Fi* has become a word. The proliferation of programs has extended down to the youngest of users and has broadened the scope of possibilities for all.

Social media continues to be adopted by the mainstream. Businesses and professionals, alike, use LinkedIn, Facebook, and Twitter as means of meeting the public and sharing information. Apps for newspapers and magazines are downloaded and are being used more exclusively in growing numbers. Storage is in clouds, and collaboration is available over multiple platforms.

Resistant or resilient, education has survived for decades by growing and adding but fundamentally remaining structurally recognizable. That day has passed. Classrooms with desks and blackboards and teachers lecturing will soon take a place in a historical photo album next to a one-room schoolhouse and a Conestoga wagon. Schools cannot be the bastions of the past. They must be the conveyors of the values and lessons of the past, but their role is to ready children for the future. And the future is not a patient partner.

● PUNCTUATED EQUILIBRIUM FOR EDUCATION

STEM possesses the first really promising potential to reenvision the educational orientation from the bottom up. From that small turn, systemic change can follow. The cause for hope comes from the sciences. About forty years ago, paleontologists Stephen Jay Gould and Niles Eldredge introduced the concept of "punctuated equilibrium" (see Gould & Eldredge, 2013) to explain the phenomenon of sudden change within a species. This model suggests that organisms can exist with very minor changes for long periods of time. Then, usually as the result of a change in an environmental or external factor, the organism reinvents itself in a very short period of time. Those who don't make the change find their survival threatened. We suggest this is where education finds itself at this moment in time.

It is already the second decade of the 21st century. Yet schools hold to ten-month school years, minutes of content instruction, bells, separate subject areas, and massive transportation systems. Too often, educational leaders wait for money to incentivize our decisions about how and what to change. Or the next policy change comes along, and educators respond. Major differences between the 20th and the 21st centuries lie in the push to national standards, the explosion of technology, the fact that resources are diminishing while a call for creativity and innovation is heightened, the global economy and the increase in STEM workplaces are all undergirded by an increasingly diverse student population with increasingly diverse needs (see Figure 1.1).

Figure 1.1   The Tipping Point

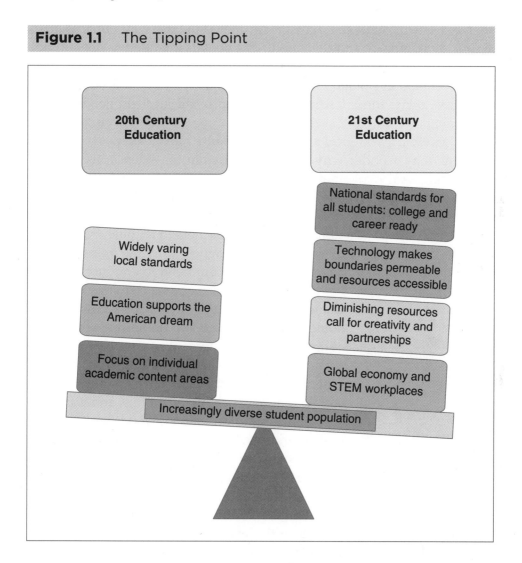

● STEM CREATES A RADICAL OPENING

The process associated with STEM requires a transition in what educators do and how they do it. It provides a vehicle to free educators from 19th and 20th century thinking, morphing how we have structured schools, teaching, and learning. STEM provides the curricular foundation for an environment that allows creativity to reenergize education and educators. That environment empowers students to be active, engaged, innovative, creative learners.

STEM is not a program that simply amplifies and accelerates the integration or focus on science, technology, engineering, and math in secondary schools. A STEM shift encourages reimagining schools, from kindergarten through the 12th grade, including the way curriculum is designed, organized, and delivered. This book is intended to open minds to an approach that has the potential to disrupt long-held traditions. Like the ox cart man in Donald Hall's famous children's story, this book invites educators to kiss those traditions on the nose and let them go. Then, we make room for new life, a new way of working together, a new way to engage students, and local answers created by local leaders.

Some may wonder why these four disciplines—science, technology, engineering, and mathematics—have become the fulcrum for an educational system shift. STEM is the first vehicle to enjoy the combined support of business leaders, government leaders, philanthropists, and communities at large. It holds exciting potential for teachers to reignite creativity in their work and to share with their colleagues, bringing reinvigorated energy back to school. It is a large enough umbrella to embrace comprehensive change. It encompasses how we learn and changes how we teach. It calls for access to technology as an unlimited and unlimiting resource. It addresses the development of 21st century skills such as critical thinking, problem solving, collaboration, agility, initiative, effective communication, social-emotional behaviors, morals, and accessing and analyzing information. It modifies how time is seen and used. It utilizes real-world problems and opportunities for authentic problem solving.

> It is not about the subject but about the learning process of inquiry, imagination, questioning, problem solving, creativity, invention, and collaboration.

The STEM shift is not a call for the elevation and focus on the subject areas of science, technology, engineering, and math, but an entire systemic shift in how learning happens. It is not about the subject but about the learning process of inquiry, imagination, questioning, problem solving, creativity, invention, and collaboration. Science is the most viable subject for inquiry. Science, technology, and engineering do not exist without mathematics. None exist without the capacity for reading and writing. Creativity is required with all inquiry and invention, so the arts are no longer for the artists but for all. STEM is an organizing principle upon which to build the interconnectedness of subjects.

Engagement and motivation connected with a topic of interest can bring a learner to a point. Application in real-world settings takes that learning into relevance and related achievement. Opportunities to apply learned information and concepts are key facets of a true STEM shift. Application of knowledge is a planned part of the curriculum and includes ever-increasing levels of difficulty and a variety of projects and problems to be investigated alone and with others.

As educators, we recognize that debate often triumphs over purpose. Here, the language and the name of the shift are caught by this propensity. Is the shift arising from STEM, STEAM, or something else? Dr. Vicki Metzgar, director of the Middle Tennessee STEM Innovation Hub, prefers the term *TEAMS*, reordering the letters to represent an underlying principle of the shift. Dr. Eli Eisenberg, senior executive director general of ORT Israel, prefers the term *iTEAMS*, identifying innovation as an essential component of the shift. Both agree a more realistic representation of the goal to integrate subjects to be active, engaged, problem solving, project based, collaborative, dynamic learning centers must include the arts. It remains to be revealed, as implementations unfold, whether the term *STEM* endures or morphs to be more inclusive.

In their book, *The Art of Possibility*, Zander and Zander (2000) remind us, "Art, after all, is about rearranging us, creating surprising juxtapositions, emotional openings, startling presences, flight paths to the eternal" (p. 3). They suggest a tightly coupled connection between scientists and artists. Inventors and entrepreneurs cannot do their work without imagination, play, and career-specific skills. The arts support imagination.

Throughout this book, we use the label *STEM*, albeit a currently debated term. STEM is an already recognizable name, with support from multiple stakeholder communities. We use it for that value, not to exclude the arts or innovation but simply to seek foundation agreement on a beginning term. This book hopes to ignite a dialogue among leaders and teachers and policymakers, and we hope to encourage some among them to unleash the potential of this giant opportunity with an entrepreneurial spirit.

In order for change to occur in the educational system that permeates practice and affects every student, a radical opening must take place. Current government policies are destroying morale, exhausting the system, and doing little to improve practice and results with too many students. The migration of schools into STEM schools and STEM districts is possible at the local level, and it can be accomplished now. Examples throughout the book support that contention.

STEM shifts practice, process, and methodology. It can release education from the constraints of time, focus training, and generate prolific partnerships. This book offers suggestions for how to make the STEM shift to a school for the 21st century. Dozens of interviews and many years of research contain examples and suggestions for the leadership steps to be taken in order to plant and grow this major change within schools and districts. It draws on the experience of STEM pioneers to reveal guideposts and land mines.

This effort is not be for the faint of heart, as it will take time, energy, training, and courage at the very least. Education is a field filled with those who care deeply about their work and who are tired of responding to mandates and following old routines. They want to think again and are standing at the ready to make the change that seems elusive.

STEM has been frequently understood as a secondary program or an elective program. This book acknowledges those as steps into a shift toward STEM but as not fully realizing the potential it possesses. **We hope to change minds and reveal the power of STEM as a simultaneous shift in the way educators work and in the way students learn.** We will share information about how children think and how leadership matters for the 21st century. All subjects are connected within STEM-themed school. Humanities and the arts are not diminished, even if their separate identities are melded. For some that may be a fear- and anger-arousing thought. But it is no more so than to say that plants can live without water. It is within the life of the plant as the arts and humanities are within the life of a STEM school.

● WHY THE STEM SHIFT NOW?

There are several reasons for this book at this time. Certainly, one reason is the growing need in the economy for workers who have STEM knowledge, skills, and dispositions. In order to sustain our economy, schools must graduate students who are prepared to work in and contribute to these fields. Secondly, all leaders must be conversant in the now trendy media affinity for STEM as the next right answer for education. This book will serve to expand STEM's meaning as a shift within education, a new process that brings authentic learning, inquiry, literacy, problem solving, communication in a variety of ways, reading and writing, and the capacity to abandon bias and recognize objectivity through all of the subjects, including the humanities and the arts. Understanding STEM is important not only for those who are interested in STEM careers but also for all of us to be able to make educated decisions about the scientific and technical issues affecting our lives (Matthews, 2007).

Most important, educators now find themselves at a threshold, one that relates to the purpose of education and locating it within the environment of this century. There are many voices telling educators what to do but few places where those who meet daily with children can invent a new way of being together in a teaching and learning community. STEM offers that; it is a choice that can be locally made and created on behalf of local and global communities.

Within the goals of education lies the business of developing independent, thinking young men and women who will be prepared to face challenges, yet unknown, in the world in which they will live and lead as adults. While no one can describe that world in detail, there is agreement that the challenges will include a type of problem solving that does not spring from a school year of listening and responding, being told what is important and what is correct. The future will require comfort with questions, returning ones and unfamiliar ones. It will rely heavily on technology for living and learning and doing business. It will demand collaboration.

Twenty-five years ago, computer scientist William Wulf coined the phrase *collaboratory*. He defined it as a "center without walls, in which the nation's researchers can perform their research without regard to physical location, interacting with colleagues, accessing instrumentation, sharing data and computational resources, [and] accessing information in digital libraries" (see Wulf, Kouzes, & Myers, 1996).

By 2003, D. L. Cogburn had refined the definition: "a collaboratory is more than an elaborate collection of information and communications technologies; it is a new networked organizational form that also includes social processes; collaboration techniques; formal and informal communication; and agreement on norms, principles, values, and rules" (2003, p. 86). Similar conceptualizations can describe the schools of the future.

The early space program offers an essential example. In 1970, *Apollo 13*, the seventh manned mission into space, met with serious and unexpected challenges. Designers, engineers, astronauts, and the flight controller were tasked with solving the problems that arose from the limited power supply, the limited time factor, and dropping temperatures in the spaceship.

The scene in the 1995 movie *Apollo 13* that tells this part of the story offers compelling evidence of the need for developing capacity at all levels and within all areas of expertise to make communication and effective collaboration possible, even in the most difficult times. It is one of the clearest demonstrations, not only of what scientists and engineers face, but also of the type of problem solving needed for and by today's students.

In the case of *Apollo 13*, there was an assumption that those skills would emerge naturally when called forth in crisis. And in that space mission, thankfully, the assumption was correct. But we cannot leave the scene's lesson behind. The particular skills required in 1970 of those working on that space mission represent the essence of the skills for the 21st century.

The lingering achievement gap, the traditions that have locked down flexibility, the need for our graduates to be career ready for a new world, and a global economy are present at education's doorstep. In the movie, Ed Harris played Gene Kranz, the NASA flight director; he said these famous lines, "Gentlemen, that's not acceptable. . . . We've never lost an American in space. We're sure as hell not going to lose one on my watch. Failure is not an option." Education has arrived at a tipping point and can take its lead from Gene Kranz.

Regardless of whether educators believe in the success of the work they have been doing, education has not kept pace with change and has not solved the problem of the achievement gap. Failure is not an option. Not on our watch. The STEM shift provides an option.

QUESTIONS AND REFLECTIONS

- All educational systems are currently living in the pressure point created by governmental mandates and the demands of an amorphous century. How is your current school or district responding to this pressure point moment?

- As you look around your workplace, where are there pockets of energy and creativity?

REFERENCES

Cogburn, D. L. (2003). HCI in the so-called developing world: What's in it for everyone. *Interactions—Winds of Change, 10*(2), 80–86.

Matthews, C. (2007). *Science, engineering, and mathematics education: Status and issues* (98-871 STM). Washington, DC: Congressional Research Service.

Rothstein, R. (2004). *Class and schools: Using social, economic, and educational reform to close the black-white achievement gap.* New York, NY: Teachers College Press.

Zander, R. S., & Zander, B. (2000). *The art of possibility: Transforming professional and personal life.* Boston, MA: Harvard Business School Press.

RESOURCES

 Access live links at http://bit.ly/TheSTEMShift.

Center for Public Education. (2009). *21st Century Skills*: http://www.centerfor publiceducation.org/Main-Menu/Policies/21st-Century/21st-century-demographics-21st-century-skills-.html

Center for Public Education. (2009). *Defining a 21st Century Education:* http://www.centerforpubliceducation.org/Main-Menu/Policies/21st-Century/Defining-a-21st-Century-Education-Full-report-PDF.pdf

Center for Public Education. (2012). *The United States of Education: The Changing Demographics of the United States and Their Schools:* http://www.centerforpubliceducation.org/You-May-Also-Be-Interested-In-landing-page-level/Organizing-a-School-YMABI/The-United-States-of-education-The-changing-demographics-of-the-United-States-and-their-schools.html

Federal Reserve Bank of Dallas. (2013). *Immigrants in the U.S. Labor Market:* http://www.dallasfed.org/assets/documents/research/papers/2013/wp1306.pdf

Forbes. (n.d.). *Top 10 Companies Hiring Foreign Workers, No. 4 IBM:* http://www.forbes.com/pictures/efei45mdli/no-4-ibm/

Gould, S. J., & Eldredge, N. (2013). *Niles Eldredge–Stephen Jay Gould in the 1960s and 1970s, and the Origin of "Punctuated Equilibria":* http://yhoo.it/1IEAKTt

Information Technology Industry Council, Partnership for a New American Economy, & U.S. Chamber of Commerce. (2012). *Help Wanted: The Role of Foreign Workers in the Innovation economy* (Report on foreign workers in STEM): http://www.renewoureconomy.org/sites/all/themes/pnae/stem-report.pdf

Kids Count Data Center (U.S. demographic information on children): http://datacenter.kidscount.org/data/tables/103-child-population-by-race?loc=1&loct=1#detailed/1/any/false/36,868,867,133,38/66,67,68,69,70,71,12,72/423,424

Ryan, C. L., & Siebens, J. (2012). *Educational Attainment in the United States: 2009* (Report P20-566): http://www.census.gov/prod/2012pubs/p20-566.pdf

Wulf, W. A., Kouzes, R. T., & Myers, J. T. (1996). *Collaboratories: Doing Science on the Internet:* https://www.cs.virginia.edu/people/faculty/pdfs/Collaboratories.pdf

2 The 21st Century Learning Environment

Science as a process is never complete. It is not a foot race, with a finish line . . . the scientists approach, pass the mark, and keep running. . . . As long as we won't commit to knowing everything, the presumption is we know nothing.

—Barbara Kingsolver, *Flight Behavior*

There are those who have been working in a STEM way for decades now and much can be learned from those 1.0 versions. But the second decade of the 21st century is already half gone, and for many, STEM is still new. Federal and state governments, business leaders, and the moral imperative are demanding change. This century's student population is rapidly becoming composed of those children whom we serve least well. They are the foundation upon which the country will be built. Science, technology, engineering, and math drive the economy and the society into an unimagined world. Twenty-first century students around the world live in a "handheld" communication world. They are connected to each other and to information 24/7. Ten-month school years, seat time in minutes of content instruction, bells, and buses are of the world that is passing away.

Innovators and entrepreneurs are purveyors of optimism in the land of possibility. They are confident and experimental and persistent. They are creative and energetic and collaborative. They are risk takers and, in Judi Neal's language, *edgewalkers* (see Resources). Educational leaders are not often described in those terms. The future is amorphous at best, but there is a confluence of forces at this moment calling educators to take a bold step, to create new educational system.

The time has come for educators to take initiative. Too much energy has been consumed by responding to mandates. Leaders have become batters in a batting cage, adept at hitting baseballs pitched at warp speed. Even the well intentioned are greedy, developing answers that don't fit the problems. In and of itself, this technological environment demands changes in teaching and learning. A consideration of STEM as an alternative begins with the reconsideration of learning theory. The science of learning should influence any change in teaching and learning.

● HOW CHILDREN LEARN

The theories of John Dewey, Jean Piaget, Maria Montessori, and Lev Vygotsky agree that children learn from doing, education should involve real-life material and experiences, and children should be encouraged into experimentation and independent thinking (Mooney, 2013, pp. 15–16). Those theories, developed in the early 20th century, continue to hold true and are validated by current research. John D. Bransford, Ann L. Brown, and Rodney R. Cocking (2000) discuss advances in developmental psychology, social and cognitive psychology, neuroscience, emerging technologies, and collaborative studies as they relate to human learning (p. 4). Bransford et al. observed

> It was not the general rule for educational systems to train people to think and read critically, to express themselves clearly and persuasively, to solve complex problems in science and mathematics. Now . . . these aspects of high literacy are required of almost everyone in order to successfully negotiate the complexities of contemporary life. (pp. 4–5)

There remains a call for learner-centered schools and classrooms where children can be problem solvers and problem generators, a call for the consideration of those hundred-year-old theories. The current challenges, however, are complicated by several factors that include, but are not limited to, the increasing communication technologies, the global nature of developed and emerging markets, a more diverse student population, an unrelenting achievement gap, the high speed and volume of new words and information, and the increasing need to be able to be prepared "to successfully negotiate the complexities of contemporary life" (p. 4).

Leading an environment capable of accomplishing that type of education is complex in and of itself. The theories stand. The ability to envision, innovate, encourage, build capacity, and reinvent schools remains unfulfilled. The problems students will face in the world are unknown. What is known is that those problems will be complex. Those working and leading and living will be called to solve those problems and adapt their knowledge, skills, and abilities to do so.

Very young children engage in solving problems with intense focus and engagement. Those very same children sometimes sit in classrooms, disengaged and drifting away, as passive recipients.

> Children are both problem solvers and problem generators: children attempt to solve problems presented to them, and they also seek novel challenges. They refine and improve their problem-solving strategies not only in the face of failure, but also by building on prior success. They persist because success and understanding are motivating in their own right. (Bransford et al., p. 112)

The design of a problem-solving learning environment requires that leaders and teachers alike understand how to use it in schools and classrooms. John Hattie and Gregory Yates (2014) argue that

> You need to know the ideas well before you start to relate them or use them in a problem solving situation. Problem solving activities impose a heavy load and can become a source of interference . . . Even when students solve the specific given problems, they may fail to acquire the underlying principles, and so fail to generalize the experience. (p. 151)

With that understanding, we do not recommend jumping ahead to implement a 21st century problem-solving environment without learning how to design those learning opportunities. No one need be surprised that a shift of this magnitude must be accompanied by intense professional development.

● 21ST CENTURY SKILLS

The Partnership for 21st Century Skills (P21; see Resources) was created in 2002 as a coalition of business leaders, educational leaders, and policymakers to "kick start a national conversation" on skills needed for this century. The unifying underlying premise is that education needed to change if it was going to produce graduates ready for successful life and work. The purpose was to impact what is taught and how it is taught in every learning environment across the nation.

P21.org has created a framework to describe the skills needed for this century (see Figure 2.1). Life and career skills, learning and creation skills, information, media and technology skills surrounding the core subjects and themes, knowledge content, expertise, and literacies are included. The concentric circles hold the support-system components for success. The rainbow of outcomes describes the knowledge, skills, and expertise students need for life and work in this century. The rainbow itself calls for a change in teaching and in the structure of schools and classrooms.

If the environmental scan demands that teaching and learning can no longer be tied only to successful past practice, then it becomes imperative to discover another organizing framework. Twenty-first century pedagogy

Figure 2.1 21st Century Student Outcomes and Support Systems

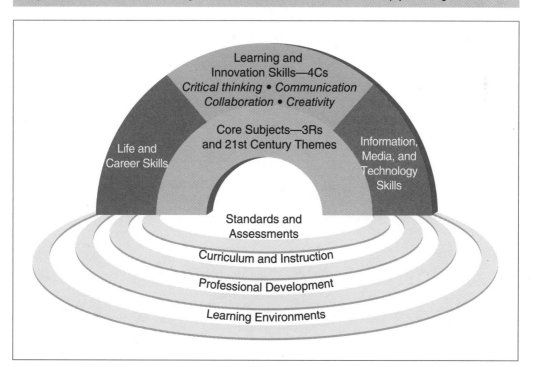

Source: Used with permission of P21 Framework for 21st Century Learning.

incorporating the multiple aspects and relationships described in Figure 2.2 must be involved. A robust and inclusive K–12 system can result.

Accompanying the understanding of the facets of the 21st century learning environment must be an understanding of what 21st century teaching must include. This Te@chThought (see Resources) diagram is a complex pedagogical model. Imagine the rainbow lies behind it. For any teacher, in any classroom, to implement both most assuredly requires a transformational systemic change that must be supported by a diverse group of leaders, teachers, parents, and community members. It must be supported by a commitment to ongoing professional development, courageous risk taking, demonstrated encouragement, political momentum, and fiscal and human capacity.

Figure 2.2 21st Century Pedagogy

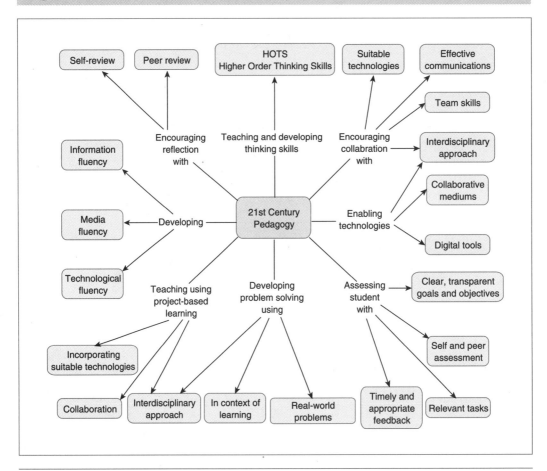

Source: Used with permission of 21st Century Student Outcomes and Support Systems.

● TECHNOLOGY BREAKTHROUGHS

Former science fiction is today's reality. Advances in technology are tachyonic in the 21st century. Commonplace communication takes place by e-mail, text messages, Skype, FaceTime, Facebook, Twitter, Pinterest, and Instagram. Each of these methods requires a different set of skills. Each allows a different way to render distance obsolete as an obstacle. The speed impacts communication, access to information, and privacy. Devices carried in pockets and briefcases reveal our location at all times. Drones can deliver packages, and cars will drive themselves. Toddlers are introduced to apps and gaming before they can read.

> Solutions to real problems can be developed and manufactured in classrooms, by students.

Traditional bastions of education cannot stand against the power of technology. It is reframing how children learn, and it can level the playing field. This world is one of programs and coding, information, and sharing. Handwriting is fading as keyboarding replaces it. Texting is creating a new language. Resource libraries are becoming virtual spaces.

Technology is breaking through bricks and mortar and redefining distance. Reading the print newspaper no longer means only holding a paper in our hands but can mean holding a phone in our hands. Communicating is no longer limited to telephone. Postcards and letters are increasingly obsolete. The capacity for communication is unparalleled, and accessibility to experts can change every classroom and any lesson. Smartphones have replaced cameras, computers, games, and calculators. Programs have become apps. Clouds once had an environmental meaning but now house information accessible from our smartphones, tablets, laptops, and desktops. Affordable 3D printers are arriving in schools. In the hands of creative teachers, they are releasing potential and improving lives. Solutions to real problems can be developed and manufactured in classrooms, by students.

Sierra Petrocelli offers just one of the many examples of what is already happening in schools. As a fifth grader in Monkton, Virginia, Sierra was challenged to decide on a topic for her science fair project. She knew she wanted to use a 3D printer in some way and intended to show pictures and explain a project. When her teacher asked her to think about how 3D printing might be able to change someone's life or change the world, Sierra decided to create an affordable prosthetic hand. She reached out to a company that makes computer models of hands. The company, in turn, sent her a tutorial. After successfully building the hand and receiving an A for her work, she is reported as having said,

"I think my favorite part is helping someone." At this writing, Sierra was producing a new hand for an eight-year-old in California. This fifth grader is learning and growing and is already in service to others (see Simon, 2014; "Eleven-Year-Old Girl," 2014).

> Our children will grow up in a world where we are all co-creators. They seem to recognize this potential intuitively and get genuinely excited about it. Our educational system desperately needs something like 3D printing to provide a more practical education that can truly engage kids. (Simon, 2014)

Our children are digitally literate and are bypassing adults in their understanding and use of these technologies. It is not a fad. Educators and psychologists are attempting to figure out how this early use of technology will change not only what but also how children learn.

● CODING

Coding is now in every facet of digital life. Programs are used in every business and industry. Programs help farmers analyze their yield and project whether they will have an abundant crop. Programs maintain financial records and banking data and track movements and habits. A smartphone does not realize its potential without apps for entertainment, for travel, and for curiosity. Every program used is the product of code. The expansion of coding has made technology and the Internet easier to use. But coding is traditionally a text-based language that requires the programmer to know, understand, and be able to use words and symbols to create the commands necessary for program development (Marji, 2014, p. 2). This made learning coding impossible for younger children.

The development of programs like Scratch and ScratchJr are helping students "program their own interactive stories and games. In the process, they learn to solve problems, design projects, and express themselves creatively on the computer" (see Resources) because they use visual programming language. With Scratch and ScratchJr, coding is accomplished by connecting graphical blocks together. MIT's Mitchel Resnick (2013), in an article in *EdSurge News*, states,

> [Children] are not just learning to code, they are coding to learn. In addition to learning mathematical and computational ideas . . . they are also learning strategies for solving problems, designing projects, and communicating ideas. These skills are useful . . . regardless of age, background, interests, or occupation. (Resnick, 2013)

Scratch and ScratchJr provide a powerful example of putting programming (coding), computational thinking, and problem solving into the hands of the youngest learners.

Through trial and error, while the children watch the results of their attempts, they receive immediate feedback while using blocks to create coding scripts. Building upon this two-dimensional experience, Carnegie Mellon's Alice: Teaching Programming through 3D Animation and Storytelling takes coding to another level and into three-dimensional design.

Scratch

Link also available at http://bit.ly/ TheSTEMShift

To read a QR code, you must have a smartphone or tablet with a camera. We recommend that you download a QR code reader app that is made specifically for your phone or tablet brand.

● THE ARTS

In his new landmark book, *Colliding Worlds: How Cutting-Edge Science Is Redefining Contemporary Art*, Arthur I. Miller (2014) refers to Ben F. Laposky, an American amateur mathematician and artist in the 1950s who produced the first electronic graphics. He describes this art as machine made, "generated through automatic processes. The role of the artist was to work with the machine and select the most appealing images, rather than creating images himself" (p. 66).

Miller posits that sixty-five years later current technology allows technicians to take an MRI scan of the brain of a person looking at a work of art. This presents the seeds of a new field called *neuroesthetics*.

Carnegie Mellon University's Alice

Link also available at http://bit.ly/ TheSTEMShift

> It is too early to know whether, as neuroscientists claim, aesthetic judgments can be explained using the laws of physics and chemistry as we know them today. And even if aesthetic judgments are totally physiological, this still doesn't help us understand what beauty is, why certain works give aesthetic pleasure and others don't. (p. 330)

According to Miller, Gerfried Stocker, artistic director of Ars Electronica, says "the age of artists working in partnership with science has passed. Rather, art is dependent on the input of science" and said, "Today art is an offspring of science and technology" (p. 305).

Peter Gelb, the Metropolitan Opera's general manager influenced a change in the presentation of opera. According to a *New York Times* article (see Wakin), in 2010 the new set for *Das Rheingold* was unveiled by creating "a new Wagner's Ring Cycle that involves a leviathan set, scenery almost entirely based on intricate computerized projections and a few age-old theatrical techniques. It is just the sort of thing, its creators say, that Wagner would have wanted: the most advanced technology in service of his opera." The motorized set weighs forty-five tons. Mezzo-soprano Stephanie Blythe reported that "the set is extremely friendly to the voice. It's a big resonating chamber. The projected scenery was simply another language for the audience to get used to. It's all very new, and it's a new experience for all of us." The set was a risk-taking collaboration among artists, engineers, architects, and programmers. It involved physics, electronics, coding, creativity, technology, and innovative thinking all working together to create, in 3D reality, what someone saw in his imagination.

These are simply a few of the art-science-technology connections that require art to be part of the STEM shift. Best explained in Miller's words,

> Artsci, the new movement I've been looking at, is more extreme yet. Not only is it science- and technology-influenced . . . but its artists use scientific and technological media . . . Today's artists often work together with scientists . . . Their work may even directly affect the work of scientists. (p. 341)

● MIXED REALITY ENVIRONMENTS

TLE
TeachLivE
Lab

Link also
available at
http://bit.ly/
TheSTEMShift

Can you imagine a virtual classroom in which teacher preparation programs have future teachers engaging avatars designed to present challenges in the classroom? It presently exists at the University of Central Florida. At this writing, in forty-two campuses across the country, the TLE TeachLivE Lab offers both preservice and inservice teachers opportunities to engage in situations and reflect on their practice in a judgment-free environment in which their skill and ability can be measured, developed, and remediated. The design of TeachLivE offers a mixed reality environment and has delivered well-researched, positive results in changing the teaching behaviors of preservice and experienced teachers alike. It, like so many other digital opportunities, is here now.

● E-LEARNING

The Internet is the highway for communication and learning. Want to know how to do anything? The lessons can be found on Web pages, YouTube videos, articles, and available courses. E-Learning has quickly become the manner in which finding information and learning how to use it happen. Universities including the likes of Harvard, Stanford, Duke, MIT, and

Carnegie Mellon have developed MOOCs (massive open online courses) that allow anyone interested to enroll and learn.

Course management systems (CMS) like Moodle or Blackboard are used in most colleges and many public schools. This allows for two pathways. One, students who need credit recovery or need to take a course that is not offered because of low enrollment, can now take a course made available in this medium. The other is a blended path in which teachers can enhance the courses they are teaching with an online component that complements the work done face-to-face. Schedules open up. Experts from anywhere can join classes as guests or as instructors. They can engage students in synchronous and asynchronous learning experiences and offer feedback online. Resources and possibilities abound.

This blended model is used in the flipped classroom. Dr. Jerry Overmyer from the University of Northern Colorado describes the flipped classroom model as encompassing

> any use of using Internet technology to leverage the learning in your classroom, so you can spend more time interacting with students instead of lecturing. This is most commonly being done using teacher created videos (aka vodcasting) that students view outside of class time. (See Resources.)

By the time students presently in K–12 schools reach college, even more advanced digital learning environments will exist. The opportunity to learn in these environments can enhance the learning taking place for students now. It can also help prepare them as e-learners of the future. In 2011, Ray McNulty wrote

> educators have to think differently—and offer something creative and new. Being able to introduce novel ideas means considering and implementing something so new that it has not been proven to work . . . if the current system isn't getting the job done, then we need to do what innovators and entrepreneurs do.

QUESTIONS AND REFLECTIONS

- How does teaching practice in my school or district demonstrate 21st century pedagogy?

- Who are the exemplars expanding the use of technology in their practice?

REFERENCES

Bransford, J. D., Brown, A. L., & Cocking, R. R. (Eds.). (2000). *How people learn: brain, mind, experience, and school*. Washington, DC: National Academy Press.

Hattie, J., & Yates, G. (2014). *Visible learning and the science of how we learn*. New York, NY: Routledge.

Marji, M. (2014). *Learn to program with Scratch: A visual introduction to programming with games, art science, and math*. San Francisco, CA: No Starch Press.

Miller, A. I. (2014). *Colliding worlds: How cutting-edge science is redefining contemporary art*. New York, NY: W. W. Norton.

Mooney, C. G. (2013). *An introduction to Dewey, Montessori, Erikson, Piaget, and Vygotsky*. St. Paul, MN: Redleaf Press.

RESOURCES

 Access live links at http://bit.ly/TheSTEMShift.

Carnegie Mellon University. *Alice: Teaching Programming Through 3D Animation and Storytelling* (object-oriented, 3D programming environment): http://www.cmu.edu/corporate/news/2007/features/alice.shtml **(QR code on page 21)**

EdSurge News. (2013). *Learn to Code, Code to Learn*: https://www.edsurge.com/n/2013-05-08-learn-to-code-code-to-learn

11-Year-Old Girl Uses Science Project to Create Prosthetic Hands for Children. (2014). (Sierra Petrocelli builds a prosthetic hand): http://kdvr.com/2014/07/15/11-year-old-girl-uses-science-project-to-create-prosthetic-hands-for-children/

McNulty, R. J. (2011). *Best Practices to Next Practices: A New Way of "Doing Business" for School Transformation*: http://teacher.scholastic.com/products/scholastic-achievement-partners/downloads/Best_Practices_To_Next_Practices.pdf

Neal, J. (n.d.). *Edgewalkers*: http://edgewalkers.org

Overmyer, J. *Flipped Learning Network*: http://www.flippedclassroom.com

Partnership for 21st Century Skills (P21). *Framework for 21st Century Learning*: http://www.p21.org/our-work/p21-framework

Scratch (programming for ages 8 to 12): http://scratch.mit.edu **(QR code on page 21)**

ScratchJr (programming for ages 5 to 7): http://www.scratchjr.org

Simon, J. (2014). *E-Nabling Sierra*: http://www.3duniverse.org/2014/05/16/e-nabling-sierra/Teach LivE (the mixed reality tool used to help teachers and leaders improve interactions with students and adults): http://teachlive.org **(QR code on page 22)**

Te@chThought: 21st Century Pedagogy: http://www.teachthought.com/technology/a-diagram-of-21st-century-pedagogy/#respond

Wakin, D. J. (2010): *The Valhalla Machine*: http://www.nytimes.com/2010/09/19/arts/music/19ring.html?pagewanted=all&_r=0

What Is STEM? (video): http://youtu.be/AlPJ48simtE

3 Clearing the Path

The truth may be puzzling. It may take some work to grapple with. It may be counterintuitive. It may contradict deeply held prejudices. It may not be consonant with what we desperately want to be true. But our preferences do not determine what's true.

—Carl Sagan

Those who yearn for innovation in their classrooms, their schools, and their districts can be given hope by the STEM initiative. They can join businesses and parents and leaders who are seeing the place for STEM. They will have to confront those who misunderstood it as just another program or an attention-grabbing acronym. In these interpretations, it will be lost as an opportunity. STEM is on the radar screen for educators, corporate executives, government leaders, and increasing numbers of parents. To some people, STEM means simply focus on the four subjects of science, technology, engineering and math, though many aren't sure about a course in engineering in the local school. Others know it as a course that integrates those four subjects. Some see it as a high school program or an annual project. Others see it as a highly integrated school-to-career system with opportunities for students to work and learn in the workplaces that require skill in these subjects.

The term is used for all or any of those things and much more. It can tip the entire educational system into a locally developed, integrated system that engages students in the world where they live as children and prepares them for the world in which they will live, do meaningful work, and be learners as adults. It prepares them actively for the careers that are already flourishing in this century and for those yet unimagined. And, it can be a vehicle for narrowing the achievement gaps existing today.

But, not everyone embraces STEM as the next right answer. How could moving the focus from the 3 Rs to science, technology, engineering, and math change achievement? Would it better prepare graduates for work or college? How could elementary teachers, whose training and orientation focused on literacy and arithmetic, switch the focus of their classrooms to integrated STEM?

Too often, an idea with immense potential is prematurely dismissed because of limiting, preexisting misconceptions. So, this chapter anticipates and clarifies concerns that surround STEM and limit its capacity to allow a new way of doing the business of teaching and learning. The following eleven mental models must be addressed for STEM to take root, make the educational experience of children a new one, and make the work of teachers exciting.

● STEM IS MORE THAN A HIGH SCHOOL PROGRAM

Across the country, schools and districts report in newspapers and on their websites about the success of high school students in their STEM-based science fair or the corporate-supported STEM high school program. To some degree, STEM is changing the high schools where courses become integrated and students excel where, previously, they did not. Perhaps, even those students who previously did not have access to these courses might enroll. This benefit cannot be overlooked. So, yes, STEM can be a program, but STEM is much more than that. When it is small, it benefits a few students, but if it is unleashed as a buildingwide or districtwide initiative, transformative moments begin to happen within the system.

● STEM ENGAGES ELEMENTARY STUDENTS

Students are best prepared for high school STEM experiences if they have already studied with teachers who have a deep understanding of the STEM concepts from the beginning of their schooling. So creating a STEM shift must begin in the early years of schooling.

Some of its concepts will be familiar to elementary teachers. Their preparation and subject-area orientation are generally more holistic. They are frequently the ones who remind us about whole child. They teach basic skills and how to learn, and in those classrooms, little ones learn what they love about school . . . or they learn to dislike it. Certainly, many middle and high school teachers can share stories about watching a student's passion ignite, but it is so much harder if those elementary years haven't sustained the child's anticipation of learning as joyful. STEM, as system shift, begins at the elementary level.

STEM captures how younger children engage with their world. They bring natural inquisitiveness and abundant curiosity to everything. They have not yet learned to restrain their propensity for questioning. They can experience wonder. They understand that in real life learning something often involves mistakes. They have learned to walk and ride bikes. They know practice makes them better. It is as if they know the scientific inquiry process innately.

There are great gaps in readiness among children entering schools. Entering kindergarten today are digital-age children. Teachers are changing instructional methodologies to incorporate who the children are and what skills they bring. These early childhood teachers may need to know more about science and math and technology than they currently do. It is not about adding the technology; it's about using it.

> Many middle and high school teachers can share stories about watching a student's passion ignite, but it is so much harder if those elementary years haven't sustained the child's anticipation of learning as joyful. STEM, as system shift, begins at the elementary level.

This is also an equity issue. If STEM can help close the achievement gap, which we contend it can, it must begin at lower grade levels. Here is where children are lost to these fields or connected to them. Reading matters, stories matter, and understanding is essential but, if weaknesses exist in science and math with many children, are educators not obligated to discover and employ an alternative?

The ability to make a difference, close the gap, and have more high school students STEM ready require a sound beginning. **So when elementary students and teachers are engaged with scientists, working collaboratively on a project or learning math through video games, they lead the system into a new place.** Middle schools and high schools must follow because the STEM-prepared students will be different and they will demand it. "No matter how many task forces are convened, how many curriculum projects are funded, or how many high-stakes tests are given, successful education ultimately comes down to the interaction and communication between a teacher and his students" (Drew, 2011, p. 77).

● STEM IS FOR ALL STUDENTS

STEM is not just for just for smart kids. Well actually, maybe now it is. Those who benefit from a STEM program at the high school level are often those who have been excelling in those subject areas in earlier grades. But there are places around the country that have demonstrated all children, at every grade level and achievement level, can benefit from learning in a STEM environment. A problem- and project-focused transsubject STEM learning environment invites all learners to become engaged in the process.

Contemporary research compels educators to develop active learners who "seek to understand complex subject matter and are better prepared to transfer what they have learned to new problems and settings" (Bransford, 2000, p. 13). All students deserve this. Within a buildingwide and districtwide STEM environment, the teaching and learning process is shifted for all.

Settings for STEM learning "vary but the nature of the task has child centeredness, extended time, well-defined outcomes and interdisciplinary mission as essential elements" (Capraro, Capraro, & Morgan, 2013, p. 88). In this environment, for example, the teacher is called upon to design opportunities for problem solving that will engage students in the learning process no matter the skill levels. **Problems can be pulled, with relevance, from the child's environment.** On collaborative projects, the knowledge, skills, and talents of each child become valued. Students who are challenged readers may be skilled artists. Students who are fluent in math skills may be challenged writers. This environment solicits all skills, if well planned, and presents each student shared successes that build confidence. As confidence is built, working on lagging skills and abilities becomes easier as part of the applied context of the learning instead of the focus of the learning.

● ADDING MORE STEM COURSES DOES NOT CREATE A STEM LEARNING ENVIRONMENT

Simply focusing on these subjects does not develop a STEM environment, and simply adding more courses in the fields doesn't either. Increasing the number of silos of study, a strategy that has been with us for over a century, can no longer be the answer. Forebears have exhausted the system's capacity to expand without letting something go. Integrating four STEM subjects to the exclusion of other subjects isn't the answer either. A basic misunderstanding is that these subjects, in isolation, even when taught by teachers who understand them deeply, will prepare students to contribute as productive adults in the 21st century.

Unfortunately, most states have graduation requirements that require schools to identify subjects separately, by course title and seat time. State-mandated tests reinforce this. Building schedules and collective bargaining agreements have translated this mentality and its related requirements into delivery systems. As resources dwindle and accountability pressures mount, creative STEM environments actually push back at those limitations and constraints.

Even if a high school offers subjects as graduation requirements that must be tested separately, flexibility can be found in how they are taught. Space must be made for collaboration with professionals from the field, either within the school environment or in the workplaces where the subject content is applied. This requires relationships with businesses, health facilities, and universities as partners in the education of our students.

● A STEM-CENTRIC SCHOOL CAN DEVELOP EMOTIONAL INTELLIGENCE

No matter the career path, emotional intelligence (EI) will be required for workplace success. In 2001, Cary Cherniss and Dan Goleman edited a book, *The Emotionally Intelligent Workplace*. In the second chapter of that book, they wrote, "in general the higher a position in an organization, the more EI mattered: for individuals in leadership positions, 85 percent of their competencies were in the EI domain" (p. 23). If intelligence quotient (IQ) is an indicator of what career field to pursue, emotional intelligence will forecast how successful a student may be in that field.

But EI is not another silo. A class in which science experiments or design models are integrated as learning projects, in which the students work in teams, research and solve problems together, and demonstrate appreciation of each other's work can also be the stage for social and emotional learning. In these applied settings and project interactions, the behavior and the lesson content are intricately connected in the active engagement and collaboration that STEM requires. In actuality, **STEM supports and magnifies the importance of EI.**

● STUDENTS WITH DISABILITIES CAN BENEFIT FROM A STEM SHIFT

An environment in which students have the opportunity to learn what they need from the very beginning, using methods of differentiation that make learning accessible, is the educator's responsibility. No longer can any student or group of students be excluded from access to this type of education. It is for **all** students.

Schools continue to struggle to provide the differentiation necessary for disabled students to graduate. The perception about the difficulty of learning science, technology, engineering, and math may have contributed to a reticence, pointing students with learning challenges away from the path that includes these subjects. Hence, we have excluded an entire population of students from being prepared for further education or employment in STEM fields.

Wounded Warriors and adaptive sports programs are challenging the mental models about disabilities, especially for the physically challenged. Educators are learning more about children on the autism spectrum who may excel in certain STEM career fields and workplaces. STEM can be truly inclusive if those in schools invest in the effort to make it so.

There is no longer an excuse to be complicit in further exacerbating the gap for our disabled students. Technology is tearing down barriers of all sorts. The goal of education in the 21st century is not simply the mastery of content knowledge or use of new technologies. It is the mastery of the learning process. Education should help turn novice learners into expert learners—individuals who want to learn, who know how to learn strategically, and who, in their own highly individual and flexible ways, are well prepared for a lifetime of learning.

● THE ECONOMY AND FUTURE STEM CAREERS DEMAND STEM IN OUR SCHOOLS

Yes, economic factors provide a strong argument for STEM. To some, it is the primary one. Certainly, the interests of businesses and corporate leaders in STEM position interested schools well for partnerships that are unique and far reaching in service to children. There are other, even more compelling reasons for STEM. **The economy and future careers are levers for a systemic shift that educators can use to serve more compelling reasons.**

STEM offers balanced, authentic opportunities for students to develop their capacities to successfully maneuver through a world in which they have no frame of reference. STEM professionals who come to classrooms, in person or virtually, are building new relationships with teachers and acquiring new respect for K–12 students. Students who spend part of their day, week, or year in a lab or design studio discover careers they might have never imagined and interests that motivate them for a lifetime.

> STEM offers balanced, authentic opportunities for students to develop their capacities to successfully maneuver through a world in which they have no frame of reference.

Not all students will pursue a STEM career. So why should we expect all of our students to have this focus? "It seems that humans have a need to solve problems . . . One of the challenges of schools is to build on children's motivation to explore, succeed, understand (Piaget, 1978) and harness it in the service of learning" (Bransford, 2000, p. 102). Whatever work one is called to do, problem solving will be part of it.

● STEM ENVIRONMENTS ARE STRENGTHENED WHEN THE ARTS ARE INCLUDED

Often arts courses are considered expendable in times of fiscal crisis; by high school, they are thought of as being for the talented few, and, except at moments of class plays, they are not interconnected with other courses in intentional ways. STEM will change that. STEM fields integrate imagination and precision. In many STEM careers, there is a reliance on photography, on graphic models, and on design elements. In his book *Creating Innovators: The Making of Young People Who Will Change the World*, Tony Wagner (2012) remarks on the conservative nature of education as a transmission system for information and knowledge from one generation to another. Without debating the value of that function, suffice it to say that it is not enough, not in this century (see Figure 3.1).

Figure 3.1 The Innovation Core

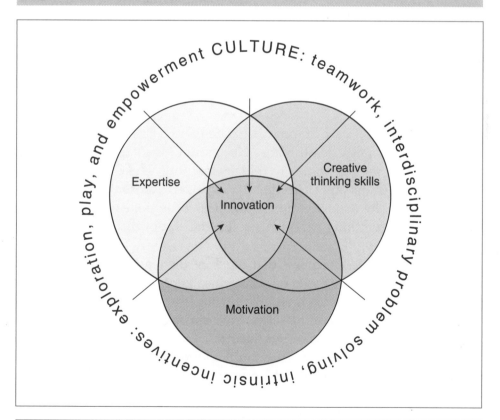

Source: Reprinted by permission from Wagner (2012).

The arts serve a STEM environment and support the development of young innovators. Those have been the domains of creative thinking and motivated learners. Add them to the mix of STEM, and a new environment emerges.

There is a new culture forming in which art, science, and technology are inseparable (Miller, 2014). Arthur I. Miller quotes a supervising technician at Pixar Animation Studios who says, "There are artists right now working in digital media, predominantly algorithmic—you might say they are procedural artists, developing algorithms and equations . . . It's artwork with a piece of code behind it" (p. 343). Like Wagner, Miller discusses the huge amount of information available in the world and highly separate, specialized subject areas as problematic in education. It leads him to examine the trend toward "interdisciplinarity" within the sciences and between sciences and the arts, leading to artsci. No matter whether one uses the name *STEM* or *STEAM* or *TEAMS*, or *iTEAMS*, STEM environments do not exclude the arts. They incorporate them as essential. Just look at this video to see how essential the arts are to even defining STEM.

> The trend toward "interdisciplinarity" within the sciences and between sciences and the arts leads to artsci.

● THE STEM SHIFT CAN HAVE A HOME IN THE COMMON CORE

The Common Core State Standards (CCSS) and STEM can coexist and, in fact, be mutually reinforcing. No matter one's position on the Common Core, the purpose is to refocus learning. The CCSS set the bar for specific learning capacities that are expected across the curriculum, in all subjects. The majority of the standards provide learning opportunities and expectations that can be located supportively in a STEM environment.

The Common Core requires that much of the reading and writing in English language arts be based on actual documents and readings from the sciences and social studies arena. Following that logic, STEM offers an imaginative path to integrating science, technology, engineering, and math with reading, writing, history, and the arts.

In fact, let's use math as an example. Historically, mathematics was taught as a series of symbols and operations with which students were expected to develop fluency. An example is the multiplication table. Students are taught multiplication tables as a memorization activity in which they can recall the correct answer with increasing speed. Stanford

University mathematician Keith Devlin (2011) observes, "You can learn to calculate with numbers without any real understanding of the underlying concepts. But applying arithmetic to things in the world, to quantities, and understanding the relationships between those quantities, requires considerable understanding of underlying concepts" (pp. 24–25).

● GIRLS DO WELL IN STEM COURSES AND WANT STEM CAREERS

A ThinkProgress.org article (2014) reported a study released by the Center for Talent Innovation that indicated that 41 percent of those graduating from engineering or science programs are women. Women are going into college with the intention of going into STEM fields. Wired.com reported that at Berkeley and Stanford girls are now outnumbering men in computer science courses where both left- and right-brained skills are essential (see Finley, 2014).

Daniel and Susan Voyer (2014) released a study, "Gender Differences in Scholastic Achievement: A Meta-Analysis," in the *Psychological Bulletin*. The results raise serious doubts about the generally accepted belief that girls are not interested in science and math. The meta-analysis asked and answered questions about the generally reported STEM gap between genders from elementary school through the university. The largest effects were observed for language courses, and the smallest gender differences were obtained in math courses. In contrast, the magnitude of gender differences was significantly smaller in mathematics, science, and social sciences courses when compared to global measures. They encourage further study to examine gender differences in school performance (see Voyer, 2014).

Looking into this issue, one finds girls like Sierra Petrocelli (see Chap. 2). Other remarkable young women are also bringing this misconception to its knees. Brittany Wenger is one of them. Brittany is a high school graduate from Out of Door Academy in Lakeland, Florida, and is now a college student at Duke University. She was included as one of *Time Magazine*'s list of "30 People Under 30 Changing the World" (see Rhodan, 2013). Why? As part of her high school work, she wrote a program to analyze needle biopsy samples that identifies malignant breast tumors with 99.9 percent accuracy. This happened because she became interested in artificial intelligence as a seventh grader. She received Google's 2012 Science Fair grand prize for the project "Global Neural Network Cloud Service for Breast Cancer." She was seventeen years old. Her contribution, and those of other girls yet to be discovered, cannot be lost. Brittany is a lesson for leaders and teachers alike. All must be watching for the untapped potential of these girls and have the STEM doors opened wide.

Young women who aspire to be scientists, doctors, technologists, engineers, mathematicians also need to become advocates for themselves, seek out courses that have been previously out of reach, and pursue their calling to new careers. They also need adult advocates who will fight for equal opportunities and equal pay for girls and women in STEM fields.

Another *Time* magazine article (see Alter, 2014) reported Dr. Ellen Kooijman, a geochemist in Stockholm, submitted a project proposal to LEGO's Idea site to develop a LEGO set that represented women in science. The awareness that LEGO was perpetuating the gender gap had been noticed by seven-year-old girl, Charlotte Benjamin, who wrote her observations that there were "more LEGO boy people and barely any LEGO girls" and observed that "all the girls did was sit at home, go to the beach, and shop, and they had no jobs, but the boys went on adventures, worked, saved people . . . even swam with sharks," (Alter, 2014). LEGO responded by accepting Kooijman's proposal. A new LEGO set, The Research Institute, including three female scientists, an astronomer with a telescope, a paleontologist with a dinosaur skeleton, and a chemist in a lab, was released in early August 2014. It sold out within weeks. The message is clear. Girls are taking their place in these new fields.

> A new LEGO set, The Research Institute, including three female scientists, an astronomer with a telescope, a paleontologist with a dinosaur skeleton, and a chemist in a lab, was released in early August 2014. It sold out within weeks.

Big Bang Theory star Mayim Bialik is another adult advocate. She is currently active in a nationwide effort to encourage girls into STEM fields. In conjunction with DeVry University, she was a 2013 representative for National HerWorld Month, a program designed to encourage girls to make a difference in the world through STEM. Bialik plays the part of a scientist on the popular TV show, but she actually holds a PhD in neuroscience from the University of California, Los Angeles.

The ability of all students, female or male, to access and succeed in courses that are rigorous and that prepare them well for college and career is an obligation of the educational community. The system can follow the lead set by corporate LEGO and by a *Big Bang* star and encourage each teacher, each leader, and each board member to become an adult advocate for girls. This talent pool cannot be excluded from the future.

● LEADERS MUST STEP UP:
STEM BACKGROUNDS OR NOT

We assert that leaders must be able to see a future and encourage others to create it with them. Remember, Moses didn't know, nor did he ever get to, the Promised Land. As the Biblical story goes, he believed it was the land where his people would thrive in the future and that it was his calling to take them there. Those who are called to lead a STEM shift can be inspired by Moses's story.

Leaders need to be convinced that STEM holds great potential for schools, for teachers and children now and in the future. So, like Moses, current and future leaders must acknowledge personal shortcomings for the task of leading schools to a new land and, simultaneously, prepare to lead the shift. Those who came into leadership positions via the humanities and social sciences route will not become scientists or mathematicians. But think of technology as Moses's staff. Technology will continue to increase capacity and support communication, invention, access to information, and the collection and analysis of data.

Leaders can also call on the others, as Moses did on Aaron, to share the leadership. **School leaders, more than leaders in other fields, are surrounded by mathematicians and scientists, as well as artists, authors, athletes, and social scientists, waiting to be invited to the futurist's table. The power and authority to include them resides in the will and heart of the leader.**

Leaders need to understand STEM, what it means, and what it offers. Then they need to examine their fears and concerns. Education, especially public education, is in its last stages without a tremendous reinvention—a seismic shift. This is a true leadership challenge. Individually, current leaders need to decide if it is the work to which they are called.

● ● ●

Each of us holds some bias or other. Misconceptions about STEM may hold back the capacity for shifting the local educational system into one designed to better prepare students for the world in which they live. While these misconceptions may not be operating within the reader, some will certainly be alive within your buildings or districts. Faculty, parents, communities, school leaders, and boards of education can encounter bias and misunderstandings as the shift process begins. Knowing these early on facilitates the implementation process.

How can we focus our professional development on elementary-level math when we have to work on our reading scores? How can we teach coding to five-year-olds? Who needs to know how to code and why? Not all kids like these subjects; what about them? How does this work in a master schedule? These questions and others will inform the investigative phase and enlarge the knowledge base as the consideration of a STEM shift moves forward.

The most insurmountable obstacle to a shift in the educational system lies within those who lead the system and those whom it has served well. The voices of cynicism, judgment, and fear express the resistance (Scharmer, 2009). They get vocalized as "I can't." "I don't know enough." "I don't have the time." "It will never happen." "It doesn't apply to me." "It won't matter." "I don't have enough money." "They won't let me do it." These oppositional arguments become positions and take on power.

These voices prevent leaders from taking hold of local schools and shifting them into the dynamic, exciting, and successful systems. They delay the progress toward preparing young people to become successful 21st century adults. They deter the cautious from considering options and looking down the road. The more cautious leader consults the rearview mirror and focuses a short distance down the road. It limits both vision and risk taking. In the rearview mirror, one can see how successful public education has been. But we do not want to miss successfully navigating the curve ahead. We forget that for this century, there is no map unless it arrives as a 3D one. Pioneer leaders, please step forward. A juncture is ahead, not unlike when wagons became trains and trains became airplanes. Schools will change. Who will lead it is the question.

QUESTIONS AND REFLECTIONS

- What misconceptions do I know exist within me?

- How is that impacting my leadership?

- How am I preparing to address misconceptions as they arise during the shift process?

REFERENCES

Bransford, J. D., Brown, A. L., & Cocking, R. R. (Eds.). (2000). *How people learn: brain, mind, experience, and school.* Washington, DC: National Academy Press.

Capraro, R. M., Capraro, M. M., & Morgan, J. R. (Eds.). (2013). *STEM project-based learning: An integrated science, technology, engineering, and mathematics (STEM) approach* (2nd ed.). Boston, MA: Sense Publishers.

Cherniss, C., & Goleman, D. (2001). *The emotionally intelligent workplace: How to select for, measure, and improve emotional intelligence in individuals, groups, and organizations.* San Francisco, CA: Jossey-Bass.

Devlin, K. (2011). *Mathematics education for a new era: Video games as a medium for learning.* Natick, MA: A. K. Peters.

Drew, D. E. (2011). *STEM the tide: Reforming science, technology, engineering, and math education in America.* Baltimore, MD: Johns Hopkins University Press.

Miller, A. I. (2014). *Colliding worlds: How cutting-edge science is redefining contemporary art.* New York, NY: W.W. Norton.

Scharmer, C. O. (2009). *Theory U: Learning from the future as it emerges.* San Francisco, CA: Berrett-Koehler.

Wagner, T. (2012). *Creating innovators: The making of young people who will change the world.* New York, NY: Scribner.

RESOURCES

 Access live links at **http://bit.ly/TheSTEMShift.**

Alter, C. (2014). *Soon There Will Be Female Scientist LEGOs*: http://time.com/2822921/soon-there-will-be-female-scientist-legos/

Finley, K. (2014). *In a First, Women Outnumber Men in Berkeley Computer Science Course*: http://www.wired.com/2014/02/berkeley-women/

Rhodan, M. (2013). *These Are the 30 People Under 30 Changing the World* (article highlighting Britney Wenger): http://ideas.time.com/2013/12/06/these-are-the-30-people-under-30-changing-the-world/slide/britney-wenger/

STEM Integration in K–12 Education (video): http://youtu.be/AlPJ48simtE **(QR code on page 32)**

ThinkProgress.org. (2014). *Women Are Leaving Science and Engineering Jobs in Droves* (women leaving STEM jobs): http://thinkprogress.org/economy/2014/02/13/3287861/women-leaving-stem-jobs/

Voyer, D., & Voyer, S. D. (2014). *Gender Differences in Scholastic Achievement: A Meta-Analysis*: http://dx.doi.org/10.1037/a0036620

4 The Achievement Gap

The have-nots in American society—the poor, the disadvantaged, and the people of color—are severely underrepresented in classrooms where mathematics and science are taught. Science education is vital for a technologically advanced society, but it is also a vehicle through which the inequalities of our society are perpetuated and exacerbated. If current trends continue, the proficiency gap in the sciences will widen between the haves and the have-nots, and this will damage our economy.

—David Drew

The U.S. Department of Education describes the achievement gap as the difference in academic performance between different ethnic groups. Even as test results improve, the gap remains. This narrow definition belies the many faces of the achievement gap in the nation. A simple truth is that there are gaps in performance—whether on standardized tests, dropout rates, graduation rates or other measures—among racial and ethnic groups, among genders, language, and socioeconomic levels of families. Ultimately, the final gap is that some groups of students leave the system more "college and career ready" than others.

The current system is preparing students for a reality that no longer exists. The American dream was realized when blue-collar jobs existed and enabled workers to rise from poverty little by little, generation by generation. Jobs of the future are not those of the last century. Technology allows access to information anywhere, anytime, and creativity in technology-based businesses is alive and well—but not for students in schools.

There is no more time to waste and no room for excuses to distort the conversation. An honest look at the gap is required and then a shift to make schools places in which children will learn and where gaps close. Even the best students, those who excel in "subjects," suffer a gap if they haven't been prepared with 21st century skills demanded for the workplace and their lives as adults.

> There is no more time to waste and no room for excuses to distort the conversation.

The gap created by poverty exists even as the children enter kindergarten and remains as a challenge for those children throughout their educational career. Gaps are associated with language differences, learning differences, and physical, emotional and mental health issues and evidence themselves in varying degrees of student engagement. The precepts of STEM project-based, problem-based integrated teaching and learning have already been seen as interceptors of these challenges that change results.

A layman's overview of our current achievement gaps reveals the following:

1. Children enter kindergarten with varied

   a. amounts of pre-school experience,

   b. amounts of experience interacting with print words and numbers,

   c. expressive (spoken) and receptive (understood as spoken) vocabulary, and

   d. experiences playing with other children and interacting with adults.

2. A majority of elementary teachers have received only minimal education about the teaching of mathematics and have received little professional development regarding mathematical reasoning and problem solving.

3. Few K–12 teachers receive training in assessment development and use in preservice education or in professional development.

4. A majority of K–12 faculties have had little time or training to develop comprehensive programs in which students can be creative and make meaning.

5. Children who enter kindergarten lacking skills in reading, writing, speaking, or mathematics will rarely be seen enrolled in higher-level courses in high school.

6. Schools with low socioeconomic populations often have severely limited resources and above-average percentages of students with high needs.

7. Many graduates require remediation once arriving in college.

8. Race, ability, language, and socioeconomic differences remain factors in the achievement gap.

Educators have vacillated between heterogeneous grouping and homogeneous grouping, pullout remedial programs and inclusive ones. They have tried addressing the needs of children living in poverty by offering breakfast and pre-school programs, and students who do not speak English with ELL programs, children who do not read well or learn mathematics easily are offered interventions, but there is no more room in the school day or year. Adding help in neat compartments is no longer an option. Tinkering has achieved all it can.

● GRADUATES MUST COMPETE GLOBALLY

Students live in local communities but are increasingly connected to a global one. There is no longer a question about whether there is a global economy. Commerce is global. Goods and services are bought and sold around the world. American companies have factories around the globe. Thomas Friedman's 2005 book, *The World Is Flat: A Brief History of the Twenty-First Century,* is no longer a groundbreaking theory. It has become an accepted reality. Tony Wagner's 2008 book, *The Global Achievement Gap: Why Even Our Best Schools Don't Teach the New Survival Skills Our Children Need—And What We Can Do About It,* points to the culture of the teaching profession as one of the culprits contributing to the failure to adapt and reinvent the system. In it he states, "Since the beginning of the twentieth century, the sort of person who was attracted to teaching as a profession was a kind of craftsman—someone who enjoyed honing a skill and greatly preferred working alone. This was also a person who valued security and continuity above challenge and change" (p. 154). Any actual shift must bring greater balance among these.

> Schools need to become breeding grounds for innovation and curiosity . . . among adults and children as well. If the resistance and skepticism can be held at bay, this is where STEM serves well.

Safety and progress are uncomfortable bedfellows. Security and change can hardly coexist. The excitement of young children cannot be stifled in schools; instead, schools need to become breeding grounds for innovation and curiosity—among adults and children as well. If the resistance and skepticism can be held at bay, this is where STEM serves well.

● SHIFT TO A LEARNING-BASED SYSTEM

What can a systemwide shift to STEM offer that is different? Douglas Thomas and John Seely Brown argue that we are living in, as their book is entitled, *A New Culture of Learning: Cultivating the Imagination for a World of Constant Change (2011)*. They describe movement from a teaching-based to a learning-based system. Not only does the learning-based approach provide environments "in which digital media provide access to a rich source of information and play," it attends to learning through engagement within the world. In the learning-based approach the goal is to "embrace what we don't know, come up with better questions about it, and continue asking those questions in order to learn more and more, both incrementally and exponentially" (pp. 37–38).

This shift is based on concepts like constructivism that are not new. Constructivist theory has its roots in the ground tilled by John Dewey, Jean Piaget, and Lev Vygotsky (see Institute for Inquiry, 1991). There is little room for platonic and all subsequent realistic views of epistemology. Tim Brown (2013), CEO and president of IDEO, describes design thinking:

> There is no such thing as knowledge "out there" independent of the knower, but only knowledge we construct for ourselves . . . learning is not understanding the "true" nature of things, nor is it, as Plato suggested, remembering dimly perceived perfect ideas . . . learning is a personal and social construction of meaning.

Information grows daily. **What is accepted as fact today is tomorrow's question and the day after's falsehood.** Human capacity to remember facts is limited. While more information is poured into the curriculum, little or nothing is taken away. And is it more room we need or the shift to a different organizing system for education? This is the essential question of the moment. Education is a publicly controlled domain with a deeply rooted belief system in a world where everything else is rapidly changing. Even those who are proponents of educational system change hold nostalgically to their own best memories of school as it was.

The 21st century requires its citizens to know and understand how to live in a world in which collaboration is a requirement and change is the norm. It also requires them to learn how to live and work with increasingly diverse people with differing skills. Tim Brown (2009), in his book *Change by Design*, chronicles

> the rise of "design thinking," a more collaborative, human-centered approach that can be used to solve a broader range of challenges.

Design thinking harnesses the power of teams to work on a wide range of complex problems in health care, education, global poverty, government—you name it.

The method of teaching and learning in STEM classrooms requires ongoing lessons in this type of collaborative behavior. As students are offered opportunities to work on projects and problems with their peers who have different abilities, talents, and skills, their capacity to be collaborative alongside those peers in the workplace is developed gradually over time. It becomes part of the learning process and not a stand-alone lesson in working together. It is embedded in practice throughout their learning careers and, thus, becomes part of their natural skill set no matter what their work or career.

Kimberly Trotter, teacher at Hattie Cotton STEM Magnet Elementary School in Nashville, reports STEM

> enhances vocabulary and critical thinking skills for elementary children. It also helps children in the areas of math and science, and allows them opportunity to see how they will use math and science in the real world. A lot of times children cannot make the connection with math and science but with STEM they are able to see, touch, connect, and engage in real world scenarios using math and science. They have the opportunity to talk with scientists and mathematicians, and can learn about all types of careers in the science and math field. (personal communication, February 2014)

Reaching the mark takes leadership and time. When a school is below the state expectations for achievement, it can be demoralizing for everyone. But once engaged in a STEM shift, teachers who make note of the continuous successes and growth and witness the students becoming more engaged as learners and experiencing more successes can counter the hidden belief that "these students" can't do it. It will build the confidence not only of the students as learners but also of the teachers making a courageous shift in their practice and their beliefs about their students.

● SENSE MAKING AND KNOWLEDGE BUILDING

There is a vast difference between understanding and memorizing. Schools have come to value a particular type of structure over the conditions required for discovery. Discovery is a messy process.

There is a good deal of evidence that learning is enhanced when teachers pay attention to the knowledge and beliefs that learners bring to a learning task, use this knowledge as a starting point for new instruction, and monitor students' changing conceptions as instruction proceeds. (Bransford et al., 2000, p. 11)

This is a nearly impossible undertaking when teacher is "teller" and student is "accumulator of knowledge." There is an operant expectation that students will graduate from high school and use the information gained to navigate their world. That requires relevancy and a capacity for transfer of knowledge.

Many approaches to instruction look equivalent when the only measure of learning is memory for facts that were specifically presented. Instructional differences become more apparent when evaluated from the perspective of how well the learning transfers to new problems and settings. (Bransford et al., 2000, p. 235)

STEM, as a systemic shift, provides the environment in which all children will have sense-making and knowledge-building opportunities as they construct, investigate, imagine, and create. Achievement gaps and learning needs in the elementary grades present different challenges than in secondary schools.

Children benefit when teachers focus on each child holistically. Children prosper with warm and sensitive teaching; integrated learning; ongoing, authentic assessment; a blend of child-guided and teacher-guided activities and the strong support and involvement of their families . . . Progress in one domain of development continues to influence and be influenced by progress in other domains; that is, development and learning do not occur in neat compartments. (Copple & Bredekamp, 2009, p. 257)

● ACHIEVEMENT FOR ALL STUDENTS

The reasons this shift can make a difference in the learning gaps are the same reasons it will change learning for all students. There are enduring understandings that are essential to decision making in designing a STEM shift. What is important for students to know and be able to do, and how will it look different in kindergarten than in high school? Scaffolding of these learning experiences can take place across grade levels as students have experiences with the same big concepts. Students in these programs, like ones in some of the most challenged schools in Nashville, Tennessee, have demonstrated raised reading and math scores in one year's time once a STEM shift has been instituted. The authentic, integrated nature of the learning engages children in the possibility that even those who are struggling have the potential to shine. STEM offers students the environment and the opportunity to demonstrate both what they are good at and the confidence to push their own limits. The environment allows for the teaching and learning needed to help them grow as learners, readers, writers, mathematical thinkers, scientists, artists, innovators, and collaborators.

> There have always been students who excel in almost any setting. They've learned to get the job done. What I am seeing with this STEM approach, which is very inspiring to me, is it has taken students who maybe were not the standout academic students in their classes and allowed them to gain their confidence and a place in the discussion. When the learning is turned around and a task is presented that requires some investigating and intuition, and philosophy beyond the student, those students with less in terms of past academic performance become active learners. They inquire, investigate, collaborate, communicate, and gain a great amount of confidence in the process . . . I think it has dramatically leveled the playing field as far as the diversity of any class is concerned. And that is something to watch!
>
> —Tyler Howe, Principal of Neil Armstrong Academy in Utah

QUESTIONS AND REFLECTIONS

- How do you lead a person into greater professional comfort with challenge and change?

- How do you build momentum for a sustainable shift?

- How might convening a design team impact the achievement gap in your school or district?

REFERENCES

Bransford, J. D., Brown, A. L., & Cocking, R. R. (Eds.). (2000). *How people learn: Brain, mind, experience, and school.* Washington, DC: National Academy Press.

Brown, T. (2009). *Change by design.* New York, NY: HarperCollins.

Copple, C., & Bredekamp, S. (Eds.). (2009). *Developmentally appropriate practice in early childhood programs: Serving children from birth through age 8* (3rd ed.). Washington, DC: National Association for the Education of Young Children.

Drew, D. E. (2011). *STEM the tide: Reforming science, technology, engineering, and math education in America.* Baltimore, MD: Johns Hopkins University Press.

Friedman, T. L. (2005). *The world is flat: A brief history of the twenty-first century.* New York, NY: Farrar, Straus and Giroux.

Thomas, D., & Brown, J. S. (2011). *A new culture of learning: cultivating the imagination for a world of constant change.* Scotts Valley, AZ: CreateSpace Independent Publishing & Authors.

Wagner, T. (2008). *The global achievement gap: why even our best schools don't teach the new survival skills our children need—and what we can do about it.* New York, NY: Basic Books.

RESOURCES

 Access live links at **http://bit.ly/TheSTEMShift.**

Brown, T. *Design Thinking: Thoughts by Tim Brown* (Web log): http://designthinking.ideo.com/?p=1165

Institute for Inquiry. (1991). *Constructivist Learning Theory*: http://www.exploratorium.edu/ifi/resources/constructivistlearning.html

5 Special Populations

One of the smartest things we can do to keep our nation globally competitive is to ensure that our science, technology, engineering and math workforce taps into America's extraordinarily diverse talent pool.

—John Holdren

Classified students, students living in poverty, and English language learners all have something in common. Although these students are not the only students facing academic challenges, they tend to be the populations that have the most disproportionate access to success in educational systems. STEM classrooms create an environment in which all learners are given the opportunity to make sense of and understand the real world. There, students remember facts because they are given relevance. With teamwork and teacher guidance, these students make the connections that allow them to generalize and apply what is learned to the next set of problems, either in the curriculum or in their world. The teachers plan the lesson, create the environment, and present the problem; they facilitate the learning opportunities and assess the students' performance. Because the students investigate, discover, and present their new learning through a variety of performance opportunities, differentiation and scaffolding can continue to be adjusted as the students' skills and knowledge grow.

● STUDENTS WITH DISABILITIES

Students with disabilities have received access to free and appropriate public education (FAPE) since 1975 when Public Law 94-142 was passed. As a result of this legislation and the work of educators in the field since then

- the majority of children with disabilities are being educated in their neighborhood schools in regular classrooms with their nondisabled peers,

- high school graduation rates and employment rates among youth with disabilities have increased, and

- postsecondary enrollments among individuals with disabilities receiving Individuals with Disabilities Education Act (IDEA) services have sharply increased. (See U.S. Department of Education, 2007)

Two recent changes are further modifying the way special needs children are included. No longer can the discrepancy between intellectual ability and achievement be used solely as the classification criteria. Now, there exists a requirement that a process based on the child's response to scientific, research-based intervention along with other alternative research-based procedures also must be used (see U.S. Department of Education Resources, n.d.).

These changes increase the need for general education teachers to know more about differentiating instruction. Teachers need a combination of subject knowledge and the skill for differentiating the curriculum. The long held belief that special education teachers were the only ones who knew best how to teach students with special needs is eroding.

General education teachers have developed knowledge and skills to identify specific learning challenges, prepare interventions, monitor and record the students' responses, and determine whether they worked or the students need another type of intervention. This process expects general education teachers to know what, for years, was considered the domain of the special education teacher. That this skill set is now shared by general and special educators supports the growth of STEM opportunities.

The advantage for all students is clear. Having educators from general education and special education work more closely together is one way in which the skills and abilities of all teachers can grow to improve the teaching and learning experiences for all students. **The STEM shift requires the collaboration of teachers and can offer the opportunity for more coteaching, which can be an asset when meeting the needs of the special needs student.**

Students with disabilities require learning environments that help them make meaning and see the relevance of what they are learning. STEM is grounded in the real world. For the student with learning difficulties, "Meaning (or relevancy) becomes the key to focus, learning, and retention" for children with special needs (Sousa, 2007, p. 13). The human brain remembers something learned based on sense and meaning. Sense making is a process by which the learner decides whether what is being learned fits with their perception of how things work. Meaning is made when the student understands how the learning relates to his or her life and requires adequate time for processing in order for the learning to move from short-term to long-term memory (Sousa, 2007, p. 14). Although true for all students, the special needs student may demand more creativity from teachers to make the learning relevant and achievable. Thinking in a STEM way makes that more natural.

Schools and families are experiencing an increase in the diagnosis of autism spectrum disorder (ASD) in children. The Centers for Disease Control (CDC) released statistics in March of 2014 that declared 1 in 68 children are identified with ASD (see Resources). A recent study titled "Science, Technology, Engineering, and Mathematics (STEM) Participation Among College Students with an Autism Spectrum Disorder" was conducted by SRI International.

Senior research analyst and author Xin Wei observed "There's a perception that people with ASDs are more likely than the general population to gravitate toward science, technology, engineering and mathematics (STEM) fields. It turns out the perception is true" (see Wei et al., 2013).

She, along with colleagues Jennifer W. Yu, Paul Shattuck, Mary McCracken, and Jose Blackorby found "The salience of STEM ability and participation among individuals with an ASD is increasing as the identified prevalence of autism continues to rise."

Integrated-problem and project-based learning at the core of STEM learning address both sense making and meaning. The opportunity to learn, both alone and as part of a cooperative group, allows students with disabilities to find their voice and their value by engaging with the learning in an active rather than passive way. In some cases, a learning strategy will be introduced, described, modeled, and practiced, and feedback will be given to help students generalize that strategy (Sousa, 2007, pp. 41–42). That is similar to the manner in which STEM-based learning takes place for all learners.

Students with disabilities also may have challenges with self-esteem. Along the way in school, they discover themselves as different from their peers. Some will have difficulty processing. Some experience difficulty with social interactions. And others have physical challenges. However, stepping into an active learning situation provides an opportunity for them to be a contributing learner. Issues of self-esteem diminish as positive learning experiences with nondisabled peers increase.

Students with learning challenges and their nondisabled peers learn best when engaged in sense making and problem solving. A STEM-learning environment invites students to synthesize and apply information learned. The STEM shift creates the best possible environment in which these opportunities can occur for all students.

● STUDENTS LIVING IN POVERTY

Students living in poverty are in a somewhat similar situation. They present a wide range of needs: physical, emotional, psychological, and academic. Teaching children who are from families with intergenerational poverty becomes even more complex. "Poverty is more than a lack of money. It becomes a way of thinking, reacting, and making decisions" (Templeton, 2011, p. 20).

President Lyndon Johnson declared a war on poverty in 1964. He called Congressional attention to "Americans living on the outskirts of hope" . . . and rallied efforts to "replace their despair with opportunity." (see Johnson, 1964) Within the next two years, Head Start programs were initiated. In those days, it was thought the war on poverty would succeed and help these students become better learners. It helped, but it wasn't enough, and it still isn't.

According to Kids Count Data, the number of children in the United States who are living in poverty grew from 18 percent in 2008 to 23 percent in 2012. This represents over sixteen million children under the age of 18. The largest percentage of them is Hispanic or Latino, but the population includes large numbers of non-Hispanic white and black or African American children as well (see Kids Count, 2014).

As these numbers have grown, so has the attempt to understand how poverty impacts learning. Lives are complicated by language barriers, the complexity of dealing with social service agencies, limited access to transportation or health care, and the limited educational backgrounds of many of the parents and caretakers of the children. Closing the gap for these students, while standards are being raised and college and career readiness becomes the goal, is difficult, complicated, and essential. Our recent past does not offer a guide for success, but STEM efforts are making a difference. As reported by SuccessfulStemEducation.org,

When students from non-mainstream backgrounds receive equitable learning opportunities, they are capable of attaining science outcomes comparable to their mainstream peers. The same is true for mathematics and, presumably, for other STEM subjects, as well. (See STEM Smartbrief)

No matter the skill level or the limits of disability or poverty, all children find a place in this process to excel. All children are invited and encouraged as contributing members of a learning opportunity. The teacher(s) are afforded the opportunity to differentiate, innovate, and intercede with individuals and small groups focusing on the skills needed to complete the tasks. The skill, whether processing information or reading and writing, becomes necessary in order to accomplish the learning mission. It provides the opportunity for the equal access to learning opportunities we truly want for all students. From Nebraska to Harlem, in public schools and charter schools, STEM is opening minds and doors.

Michael Steele is principal at Stratford STEM Magnet High School, also in Nashville. The student population has a 92 percent free and reduced lunch rate and is 77 percent minority. Steele identified a limiting belief that can arise among educators who work with students living in poverty. It is that they will not able to succeed or do the hard work involved in STEM fields.

As far as whether they can do the work or not, we've proven otherwise because the rigors that we've changed at the school have gone up tenfold and they're doing the work, they're getting academic scholarships, they're being accepted at college . . . So, we've got kids that are making the dean's list at Belmont University and things like that. So, as far as evidence, our ACT scores have gone up each year. Our academic achievement has gone up in math and in literacy. Sometimes we don't make the mark that the state wants but we always have growth . . . When you have a school like Stratford that five years ago that was one of the worst performing schools in the state of Tennessee and was on a list of the most dangerous schools in Tennessee to not be in the category anymore, to not be on that dangerous list anymore, is a huge accomplishment for our students. (personal communication, May 21, 2014)

● ENGLISH LANGUAGE LEARNERS

English language learners also have access to mandated services. Like all students, they present a wide range of strengths and challenges. Some come from countries in which they had excellent schooling, and others have come with little or no previous schooling. The goal is to support them so they can reach current standards as quickly as possible and graduate with their age peers.

In a beginning-level ELL class, there could be a child from one country whose age places him in the fourth grade. That particular child may not have attended school on a regular basis in his country of origin and be placed in this class with a child from another country who attended school regularly and was a top student. Both have yet to master English, but they are not at the same entry point. Considering the data, this is a growing challenge. The number of school-age children whose home language is other than English doubled between 1980 and 2009 (see Lombardi, 2004).

In 2013 the U.S. Department of Education reported that

In 2010–11, states in the West had the highest percentages of ELL students in their public schools. In 8 states, 10 percent or more of public school students were English language learners—Oregon, Hawaii, Alaska, Colorado, Texas, New Mexico, Nevada, and California (California data were imputed from 2009–10 data)—with ELL students constituting 29 percent of public school enrollment in California. Thirteen states and the District of Columbia had percentages of ELL public school enrollment between 6 and 9.9 percent. In addition to the District of Columbia, these states were Oklahoma, Arkansas, Massachusetts, Nebraska, North Carolina, Virginia, Arizona, Utah, New York, Kansas, Illinois, Washington, and Florida. The percentage of ELL students in public schools was less than 3 percent in 13 states; this percentage was between 3 and 5.9 percent in 16 states. The percentage of ELL students in public schools was higher in 2010–11 than in 2009–10 in just over half of the states (28 states), with the largest increase in percentage points occurring in Nevada (3 percentage points) and the largest decrease in percentage points occurring in Minnesota (2 percentage points).

In 2011 and in all previous assessment years since 2002, the National Assessment of Educational Progress (NAEP) reading scale scores for non-ELL 4th and 8th graders were higher than their ELL peers' scores. This disparity is known as an achievement gap—in NAEP reading scores, the achievement gap is seen by the differences between the average scores of two student subgroups on the standardized assessment. (Fast Facts, 2013)

These three subpopulations—students with disabilities, students living in poverty, and English language learners—are not totally distinct

Students who are not typically found engaged in or attracted to these subjects, when introduced to these learning environments, find themselves active and engaged in investigation, discovery, and application (see Figure 5.1).

groups. There is much overlap. When they are engaged in the guided social interactions required in a STEM-based learning experience, language use is encouraged; the motivation for conversation is natural, especially with their peers.

Identity, engagement, and motivation are important factors in improving adolescent literacy for native and nonnative English-speaking teens alike . . . they usually view peer interaction and collaborative literacy positively. Perceptions of themselves as, for instance, good versus slow readers, influence their motivation. (See Short & Fitzsimmons, 2007)

Direct instruction does have a place in the second language acquisition for these students, as it does for all students. Nevertheless, learning experiences that encourage acquisition of language and subject matter can be increased through the use of project- and problem-based learning opportunities in STEM classrooms.

Figure 5.1 All Student Learning

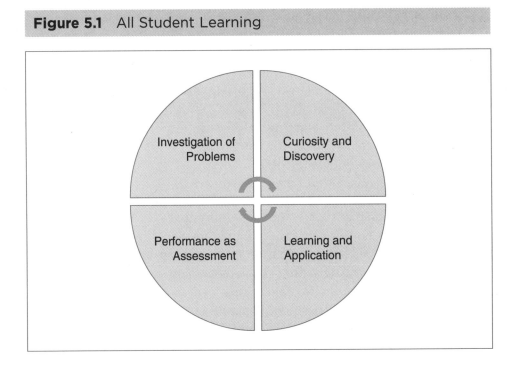

● ALL STUDENTS

All 21st century students deserve the opportunity to be prepared for college and careers. Careers in the STEM fields and the advantages for students of a STEM learning environment argue for the STEM shift. All students are offered access to learning opportunities in these environments because science, technology, engineering, and math lend themselves to project-based learning and improve critical thinking and communication skills (Capraro, Capraro, & Morgan, 2013, p. 122).

QUESTIONS AND REFLECTIONS

- How do we create learning environments in which limitations and differences do not become excluding?

- Where do we celebrate "I can't" becoming "I can" with enthusiasm equal to the highest successes?

- Who speaks for special populations at the STEM-shift table?

REFERENCES

Capraro, R. M., Capraro, M. M., & Morgan, J. R. (Eds.). (2013). *STEM project-based learning: An integrated science, technology, engineering, and mathematics (STEM) approach* (2nd ed.). Boston, MA: Sense Publishers.

Sousa, D. A. (2007). *How the special needs brain learns* (2nd ed.). Thousand Oaks, CA: Corwin.

Templeton, B. L. (2011). *Understanding poverty in the classroom: Changing perceptions for student success.* Lanham, MD: Rowman & Littlefield.

RESOURCES

 Access live links at http://bit.ly/TheSTEMShift.

Centers for Disease Control. (2014). *CDC Estimates 1 in 68 Children Has Been Identified With Autism Spectrum Disorder:* http://www.cdc.gov/media/releases/2014/p0327-autism-spectrum-disorder.html

Johnson, L. B. (1964). *91st Annual Message to the Congress on the State of the Union* (January 8, 1964): http://www.presidency.ucsb.edu/ws/?pid=26787

Kids Count Data. (2014). *Child Population by Race*: http://datacenter.kidscount
.org/data/tables/103-child-population-by-race?loc=1&loct=1#detailed/1/
any/false/36,868,867,133,38/66,67,68,69,70,71,12,72/423,424

Lombardi, J. (2004). *Practical Ways Brain-Based Research Applies to ESL Learners* (ESL
learning research): http://iteslj.org/Articles/Lombardi-BrainResearch

National Center for Education Statistics. (n.d.). *Fast Facts: English Language
Learners*. U.S. Department of Education, Institute of Education Sciences:
https://nces.ed.gov/fastfacts/display.asp?id=96

Short, D. J., & Fitzsimmons, S. (2007). *Double the Work: Challenges and Solutions to
Acquiring Language and Academic Literacy for Adolescent English Language Learners*:
http://carnegie.org/fileadmin/Media/Publications/PDF/DoubletheWork.pdf

STEM Smartbrief: Raising the Bar: Increasing Achievement for All Students: http://
successfulstemeducation.org/resources/raising-bar-increasing-stem-
achievement-all-students

U.S. Department of Education. (2007). *History: Twenty-Five Years of Progress in
Educating Children With Disabilities Through IDEA*: http://www2.ed.gov/
policy/speced/leg/idea/history.html

U.S. Department of Education, Office of Special Education Programs. (n.d.).
Building the Legacy: IDEA 2004: http://idea.ed.gov/explore/view/p/,root,
dynamic,TopicalBrief,23

U.S. Department of Education, Office of Special Education Programs, IDEA:
http://idea.ed.gov/explore/home

Wei, X., Yu, J. W., Shattuck, P., McCracken, M., & Blackorby, J. (2013). *Science,
Technology, Engineering, and Mathematics (STEM) Participation Among College
Students With an Autism Spectrum Disorder*: http://www.ncbi.nlm.nih.gov/
pmc/articles/PMC3620841/

CONCLUDING THOUGHTS ON PART I

In **Part I** we have introduced the concept of a STEM shift. It may have been new to you or reinforcing of your existing thinking. Either way, if you are convinced that a STEM shift has value for your school or district, read on. **Part II** contains the lived stories of others like Kim Trotter and Michael Steele, who have experience in the shifting process.

In the Preface, we described the stages of the STEM shift process and the expanding characteristics as the shift becomes more expansive. The figure below recaptures the titles of the stages. The voices in Part II will come from each stage. Before moving on, we invite you to take a mental walk around your school or district. As you do, discover and assess where your entry point may be—remembering you can enter in any stage.

STEM Shift Stages

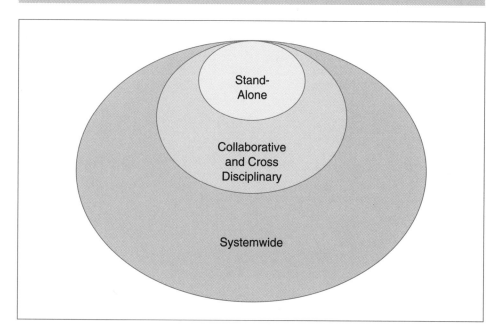

Part II

Shifting

Trust Line in the Shifting Process

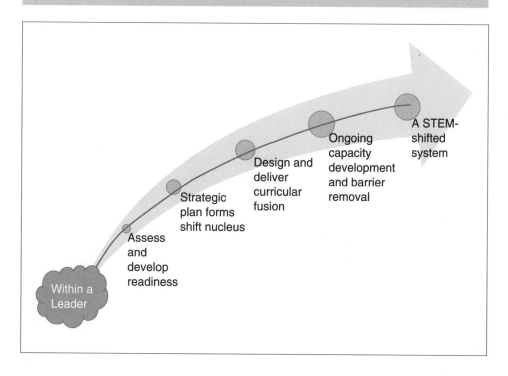

6 The Shift Begins Within a Leader

Many of the world's great movements, of thought and action, have flowed from the work of a single man. A young monk began the Protestant reformation, a young general extended an empire from Macedonia to the borders of the earth, and a young woman reclaimed the territory of France. It was a young Italian explorer who discovered the New World, and 32-year-old Thomas Jefferson who proclaimed that all men are created equal. "Give me a place to stand," said Archimedes, "and I will move the world." These men moved the world, and so can we all.

—Robert Kennedy

Schools have been fundamentally the same for over a century. Buildings are recognizable, outside and inside. Meanwhile, the world in which schools exist is alive in a life process, as all living things are. Actual change within schools has occurred in painfully small increments. **But beware or be excited, STEM holds the potential to change the system itself and, at the moment, it is not being mandated.** It awaits the choice of leaders courageous enough to enter the transformative stage of chrysalis, frightening as it is, believing in the powerful and exciting educational system that will emerge.

● LEADERSHIP RESERVOIRS

No school district, or school for that matter, can prepare for a systemic change without a profound and abiding understanding among that system's leaders. Leaders may have vision but not the skill to guide a system through a fundamental shift. Others may be very knowledgeable but not have the capacity to generate uncoerced followership. There are four reservoirs upon which leaders must draw if they chose to begin this process (see Figure 6.1).

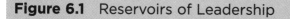

Figure 6.1   Reservoirs of Leadership

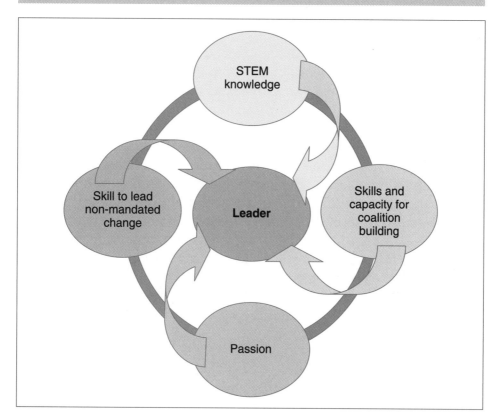

How and where does the leader learn to pull from those reservoirs at the right moment and in the right degree? It is the untaught art of leadership. Some come to this place of being the leader more intuitively, but for most, it takes experience, intuition, and practice. It takes careful observing and listening to the stories of those who have mastered the art.

Valerie Brown and Kirsten Olson (2014), authors of the well-researched new book *The Mindful School Leader*, introduced us to the work of Linda Stone. She is the former Microsoft executive who is now the force behind The Attention Project. She argues that "we are fully responsible for how we choose to use this extraordinary tool" of attention. Her description of "continuous partial attention" aptly describes the state of life for leaders. This state often derails or undermines the flow between and among the reservoir pools. So does stress. Yet leaders who aspire to undertake a STEM shift must not become subject to the diversions that disrupt and distract them. It is attention that allows the leader to stay open for the perfect blending between and among the reservoir pools.

As the title of their book suggests, Brown and Olson are proponents of mindfulness practice for school leaders. Interestingly to us, they share some of the same conclusions that we reached from a STEM perspective.

> As learning is a lifelong process, school leaders can profit from understanding the science of the brain in service of educational leadership. Increasingly, school leaders are called to take in "the big picture" and to reexamine some of the assumptions (now outdated) that have guided instructional design in conventional educational settings for decades. The kind of big-picture engagement . . . is now borne out by neuroscientific data. . . . This global picture means understanding the human implications of education not only in terms of curricular and pedagogical theory but also in terms of the very real health and vitality of the basic work of teaching and learning: listening, observing, focusing, reflecting, and so forth . . . The success of our schools will not be built on antiquated models of learning, but instead will emerge from an approach that honors the whole person. The mind and the intellect cannot be separated from the heart and the spirit. (pp. 56–57)

Passion appears at the bottom of the Figure 6.1. It is the heart source. **It is an undeniable authentic calling, beyond commitment, deeper than choice.** Passion distinguishes a leader from the political charge of being chameleonic. It is energy, the fire that burns and the coal that simmers. Within the leader, it congeals around purpose and causes action. Like a chemical catalyst, it changes the nature of the other reservoirs, making them stronger, giving them voice and words that speak to the heads and hearts of followers and partners. Passion is rational and heartfelt.

STEM leaders begin the shift by turning their minds to wonder and then encouraging others to do the same. "Without the capacity for wonder, we will most likely remain stuck in the prison of our mental constructs" (Scharmer, 2007, p. 134).

In the research supporting this book, many leaders reported that the initial vision was not theirs. Instead, someone close to them—a teacher, a parent, or a community member—knew enough about STEM to lead them into the vision. While the passion reservoir is solely within the leader, the STEM knowledge reservoir contains all those who know STEM and can see the vision. The leaders who will take STEM from a project or a program to a school or a system, however, must understand the meaning and consequences of a STEM shift in their district.

To do this requires the leader, himself or herself, to become Bridges's (2009) trapeze artist, willing to release the past, stepping into disequilibrium, trusting in the future yet to come. C. Otto Scharmer (2009) calls this letting go and letting come. At this writing, a STEM shift is a non-mandated change. Leaders will not be calling out the level of government requiring this disruption or naming the law or formidable consequence as a lever to force internal and external acceptance. This is local choice. The breadth of skill necessary to lead non-mandated change is greater than for a mandated one. The skill sets in the change reservoir are systems ones and human ones. **Leaders, even when fully engaged as change leaders, are not immune from fear or vulnerability.** In fact, it is the leader's courage in the presence of these emotions that generates leader credibility. These feelings are also the ones that allow for compassion, creating in the leader capacity to hold out a hand, not a fist, to those who are reticent to join the shift.

When teachers and students begin to work together across disciplines on real problems, with different time schedules and performance assessments, when technologies are integral not peripheral, when programs are being created as well as used, when the elementary teacher can't tell if the lesson is science or reading because it is both, when the science teacher and the art teacher need each other, when the math teacher and the drama coach collaborate on set design, when the walls don't matter and the doors open up, there is a new school life emerging. These environments need those who live and lead change. Tony Wagner (2012) describes disruptive innovation as distinct from incremental innovation. A STEM shift can begin incrementally, but a STEM building or system shift will be a disruptive innovation.

> The right issue and the right time have come together.

No leader, regardless of skill or experience or passion, can lead a system shift singlehandedly. **The good news is that STEM allies are easy to find.** There will be some within the system and many outside it. The skill is to find them and connect them and make them an energized coalition for change. The leader taps into the coalition reservoir. The old wisdom was that change had no constituency but the status quo had a strong one. Unlike a mandated change in which an external hammer of some sort is the motivation for the unwilling to comply, STEM changes are locally

generated and emerge on local timelines. Teachers, school leaders, superintendents, board members, parents, members of the community, and
representatives of local colleges and businesses can convene at the table
begin the process. The right issue and the right time have come together.

● NOT WITHOUT HEART

This shift will not be complete without addressing issues of the heart. For
now, let us agree that a new day is dawning and, as on any day, it presents
the gift of choice, to choose how we will live it. The environment will be
one of loss—loss of the old way, of security. Basic Maslovian needs arise,
and space for grief needs to be held. The voices of fear, cynicism, and judgment emerge to counter the forward movement of the shift (Scharmer,
2009). In other instances, leaders, too, have fallen prey to them. When that
happens, inevitably, it will happen within the faculty as well. There it
destroys classrooms and hurts children. Only one antidote exists; it is
found in the leader's heart, and it can be a contagious remedy.

Heart and courage are hard to come by. And they are interconnected.
The root of the word *courage* from both its French and Latin origins is the
word that means "heart." There are seldom courses or meetings in which
leaders discuss the growth of the heart or evidence of leading by heart.
Ask children what is in their heart, and you will likely hear a combination
of facts, feelings, and fantasy. Ask adults, and you will likely hear what
they know or think. Let us learn from the children.

Parker Palmer's (2011) definition of *heart* is the place where intellect,
emotion, imagination, and intuition are integrated (p. 6). There, one dis-
covers the passion and power of wholeheartedness and the pain and suffering of being brokenhearted. In the difference lies the source for action
or for despair. Since there is no more time for despair, leaders must become
wholehearted. No one will follow a leader whose lack of compassion insu-
lates and separates, at least not into new territory.

Others are now brokenhearted and suffering loss and fatigue. That
needs to be acknowledged but cannot become a place to settle. Teachers and
leaders alike yearn to be called back to the active front of making a difference empowered with efficacy. It is the leader's role to bring others along, to
help them see this as the beginning point. **A wholehearted leader can share
insight, reveal wisdom, and release courage.** It is also when we can ignite
hope. No one will follow a leader whose words are not enlivened by hope-
fulness. Hope is in short supply these days, and hope is the essential water
for sustaining and motivating those who enter the STEM shift. It breeds
excitement, the kind children and the adults who work with them deserve.

Bill Moyers (2008) wrote that "without hope we lose the talent and drive to cooperate in shaping our destiny." It is a violation of a leader's responsibility to ignore or destroy the talent existing in schools; a leader or a school community should not abandon its future to the control of others. The efficacy to create a community's future resides in the hands of its leaders, even if it is influenced by mandates put in place by others.

Right now, every educator needs hope. Each leader is holding the tension between loss and possibility, between what is and what might be, between the terrain well traveled and the unexplored region where vision and heart call us as bushwhackers. While holding that tension in truth and with honesty, leaders ask others to follow, to openly trust, to become cocreators, and to launch off from the comfortable into the journey. Without heart, hopes to create schools that offer children the best of 21st century learning will be dashed.

Michael Steele, principal of Stratford STEM Magnet High School in Nashville, Tennessee, speaks about heart:

> When we look at children we need to look and to learn how— mostly in our heart—to look at students differently. Every child is different, you know. We've got to understand the culture. We've got to understand children that are coming from abusive homes, some with different styles of learning . . . But it has to be a heart change, it cannot be a program or a grant, or this book, or this program, or this training, this PD, it's got to be a change of the heart. And so I would encourage leaders to evaluate and learn more about their level of emotional intelligence so there can be an openness to a heart change. (personal communication, May 21, 2014)

And so we locate this chapter about the leader as the first one in Part II. Why? **In every place where we found the STEM shift taking root, there was a leader at its core**—someone, like Michael Steele who had asked himself hard questions and found the courage to change his own heart, first.

● VISIONING

In a STEM shift, visioning becomes an ongoing process. Those who ascribe to the operating principle that the vision is a description of the horizon will understand that as one proceeds on the journey the view of the horizon changes. The inner and political challenge for a leader is to not confuse the North Star with the horizon.

Creating an individual vision, one that is clear and concise and moti-vating, is a difficult task. Creating a vision, collectively, for a system, is even more complex. Multiple layers of personal judgment and bias and experience enter the fray. But neither the personal one nor the organiza-tional one will serve well if not operationalized with integrity. The leader's role in making vision transparent and meaningful requires staying closely engaged with the shift process, framing the guiding questions and inviting more and more partners into the process.

We might be convinced that our economy will continue to be based on science, technology, engineering, and math. We may be committed to edu-cating children to be college and career ready. We may believe that we are preparing children for a world we don't yet know, but we think children will live and work in a more diverse society. Though that world may be collaborative or competitive, developing skill and appreciation for differ-ences and teamwork in school provides good foundation.

We do know that innovation in technology, science, and medicine is continual and escalating in speed. We might also agree that those who can solve problems are best suited to finding paths through disintegrating structures, systems, and ideas. Might we agree, then, that problem solving is as important as, if not more important than, the specific problem solved?

A leader's progress and success in leading this STEM shift will sit firmly in his or her own capacity to see the frontier as an exciting, challeng-ing, and necessary place to be. Then that leader must want to show and tell others. Finally, the leader can invite the others who are most in wonder to come out to the edge and see it also. This begins a shift. For educators, **it is a shift that transcends subjects yet excludes no subject.** It possesses the capacity to be as large and sweeping as we want to make it. Success, down the line, will be based on each graduate's capacity to read, think, understand, write, work well with others, solve problems alone and with others, care deeply about the planet and one's neighbor and contribute meaningfully to sustain the economy and our democracy.

> A leader's progress and success in leading this STEM shift will sit firmly in his or her own capacity to see the frontier as an exciting, challenging, and necessary place to be.

● DON'T FORGET

Getting lost in the "how" and the checklists stifles the heart. Without heart, energy is lost, interest wanes, and the journey is interrupted, or worse, ended. It is heart, passion, and purpose on a higher ground than testing that gives STEM leaders the energy and focus, the courage to do what it takes to stay the course and lead a transition through to a fundamental shift successfully.

REFERENCES

Bridges, W. (2009). *Managing transitions: Making the most of change* (3rd ed.). Philadelphia, PA: Da Capo Press.

Brown, V., & Olson, K. (2014). *The mindful school leader: Practices to transform your leadership and school.* Thousand Oaks, CA: Corwin.

Palmer, P. J. (2011). *Healing the heart of democracy.* San Francisco, CA: Jossey-Bass.

Scharmer, C. O. (2007). *Theory U: Learning from the future as it emerges.* Cambridge, MA: Society for Organizational Learning.

Scharmer, C. O. (2009). *Theory U: Learning from the future as it emerges.* San Francisco, CA: Berrett-Koehler.

Wagner, T. (2012). *Creating innovators: The making of young people who will change the world.* New York, NY: Scribner.

RESOURCES

 Access live links at http://bit.ly/TheSTEMShift.

Moyers, B. (2008). *Democracy in America Is a Series of Narrow Escapes, and We May Be Running Out of Luck:* http://www.alternet.org/story/85521/moyers%3A_%27democracy_in_america_is_a_series_of_narrow_escapes%2C_and_we_may_be_running_out_of_luck%27

7 Entering the STEM Shift

Before you disturb the system in any way, watch how it behaves. If it's a piece of music or a whitewater rapid or a fluctuation in a commodity price, study its beat.

—Donella Meadows

Deciding to be a school in a STEM shift can be what Jim Collins called a "Big Hairy Audacious Goal." Although that idea is now decades old, it comes to mind here as STEM leaders attempt to create goals so "clear, compelling, and imaginative" that they drive the change process. That BHAG must be contextualized within the acknowledgment that schools are centers of learning for children, that STEM serves and focuses that purpose, that talent and opportunities already exist for developing toward the goal, and why STEM is selected for the local community.

The beginning of the process will set the stage for the shift. Skipping ahead to "Let's do STEM" can be the first step toward a failed reform. There are questions that help shape the path that each school and district will benefit from asking:

- Why should we create a STEM system?
- What is our vision for a 21st century school system?
- Will STEM mean that the arts and social sciences will be left out of the future?
- Does a school or district need to choose the STEAM name to demonstrate that arts remain essential?
- Will all students participate as beneficiaries or just the "chosen few"?

Multiple layers of personal judgment and bias enter the fray and must be revealed and addressed.

● INFORMED INDIVIDUAL DESIGNS

The complex nature of schools and the differences in circumstance require that a STEM shift be individually designed by each school system. Research conducted for this book tells a story and confirms there is no one route followed by those who have entered the STEM waters. Whether the movement begins from a group of interested teachers, parents, students, administrators, or business leaders, or is encouraged by a state education department, a common understanding of what STEM will mean locally must be forged.

In addition to examples of systemwide shifts, the chapters that follow include examples of some who have a STEM project in a grade level, a club, or a course at a high school. In a systemwide implementation, those engaged in these microcosms of the shift can become an inner constituency to support a more holistic shift. They can offer their experience as a guide. But leaders who are not fully engaged can stifle or halt any innovative shift.

Examination of plans and successful experiences of those who are a few years ahead in implementation is informative, but replication of those models creates less local investment. Often the models chosen end up falling short of the goal and become unsustainable without external funding. Or they become an appendage to the traditional system, which continues uninterrupted on its 20th century course. In that case, a few select students benefit, but the others do not. "Trickle in" as an approach has not yet reformed our system. There are general considerations that can be taken from other programs that can help drive decisions.

> "Trickle in" as an approach has not reformed our system.

The beginning point is an interest in, or commitment to, one or all of the following:

- A shift to subject integration and problem-based or project-based learning in all subjects

- A desire to lead into currently non-mandated reform with local initiative

- An openness to increasing collaboration between classroom teachers and STEM professionals in the field

- A curiosity about the possibilities technology presents for new ways of learning and for energizing the teaching and learning relationship

Once the study of existing programs begins and the visits and conversations unfold, ideas about the many faces of a STEM shift reveal themselves. It is critical that these conversations and ideas be captured because what may seem unrealistic today may become possible tomorrow. It is important that no thought be left behind (see Figure 7.1).

Figure 7.1 Beginning the Discussion

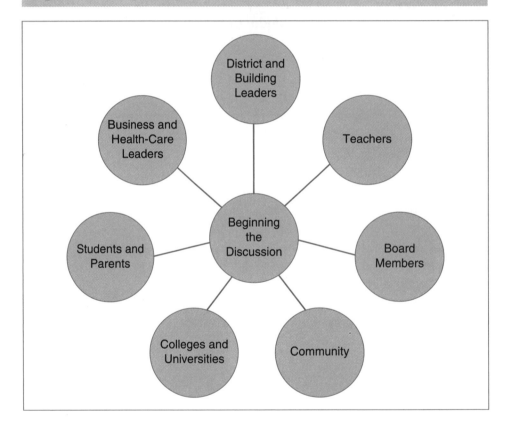

It is also very important that questions are clarified as the designers of the vision find their way to articulate it. Questions make clear the purpose and direction. "Just as scientists clarify and crystalize a research question before beginning an experiment, so too must designers clarify their task" (Scharmer, 2007, p. 132).

In East Syracuse Minoa, a team went from being the STEM team to the STEAM team to include the arts, and then the STREAM team, with the R representing the research component of the work. The evidence that this is an all-inclusive shift lies in the ever-expanding acronym. Businesses, universities, and governmental agencies are all anxious to support the development of STEM programs. Their attention has brought schools more private and public support than has been previously extended.

Corporate leaders and those in higher education and government coalesce around the belief that the development of STEM "programs" will prepare workers for the future, ensuring that the United States will have a capable workforce. From educators' perspectives, the value of the STEM shift is all of that and more. It addresses the challenges we are facing in teaching and learning—the work we are facing daily.

Engaging with problems and projects, developing inquiry skills along with the integrated transsubject topics for study are good for the best and the brightest as well as the most challenged learners. All students can flourish in an environment that allows them to investigate, collaborate, solve problems, and perform their learning. The words we choose to communicate the shift and the words we use to describe it can and will make a difference in its acceptance.

Helen M. Luke (2000), Jungian analyst and author, wrote about the difference a quarter turn can make in how an object is seen and understood. In a similar context, C. Otto Scharmer (2007) proposes three movements that achieve a turn and actually help us see more clearly from the beginning: "(1) clarify question and intent, (2) move into the contexts that matter, and (3) suspend judgment and connect to wonder" (p. 313). This will lead to opening minds, hearts, and wills for visioning what can be possible and beginning to create prototypes.

Progress and success in leading this STEM shift will sit firmly in one's capacity to see the frontier clearly and communicate the understanding of it. This is a shift that transcends subjects, excludes no subject, and has the capacity to be as large and sweeping as we can make it. Success will be based on each graduate's capacity to read, think, understand, write, work well with others, solve problems alone and with others, and care deeply about the planet and his or her neighbor.

Change is easiest when there are no choices and survival is at stake. But even then, it presents challenges. Nowhere is change more difficult than in schools where school rituals and traditions are part of the fabric of the entire community. The more members of the community who enter the change process with the leaders, the further ahead one is. If an organization believes it does not need to change, the starting place itself will be more painful. Asking those who have a tight grip on the familiar and the secure to let go will exhilarate some and terrify others. The leader might need the skills and the good will of both sets of potential partners.

● ASSESSING READINESS

If you are ready to begin transitioning into a Stage II or III STEM shift, you have been studying and talking with the school community about how children learn and how subjects are interconnected in the real world. You have an educated board, superintendent, principal(s) and faculty(ies). They have read and seen, discussed, and understood how STEM can

- reach more learners;
- impact student engagement and achievement; and
- offer an integrated curriculum that presents opportunities for all children to learn and solve problems using information from various "subjects," including the arts and humanities, science, technology, engineering, and math.

In addition, a critical mass agree that "all students can master science and math and can be prepared for America's future global economic challenges" (Drew, 2011, p. 2) They understand that "all students" include girls, minority students, those living in poverty, those for whom English is a second language, and those with learning disabilities. There is also agreement that complexity of learning can increase as students develop their abilities in a STEM-shifted system. "Competent mathematics and science education is critical if the United States is to compete in a globalized, high-tech information economy" (p. 13).

Maintaining an open, transparent process as the learning takes place and shifting begins helps keep everyone informed. Those involved, those interested, those fearful, those watchful, all can be kept in the loop. They will feel more welcomed into the process, if even only as a voyeur. Those with objections, who feel heard, are important to this process because they are the ones who will expose the speed bumps, identify stuck points, and shine the light on flaws or weaknesses in our thinking. The development and communication of the vision are the key to opening the door for the plan to follow.

● PARTNERSHIPS

Community and school readiness will be accelerated by external partnerships, especially if those partners are investing in the development of new relationships with the schools, the students, and the teachers. Establishing those partnerships is essential to the shift. Other resources already exist and are there for the asking. Examples from our research will highlight partnerships with universities, with businesses, and with the medical community.

As a result of Race to the Top's efforts to provide incentives, STEM Networks and STEM Hubs were formed in many states. For example, Battelle manages STEM networks in both Ohio and Tennessee. STEM networks and STEM Hubs have developed across the United States to promote the growth of STEM education and practices, but if your state has neither, information from the websites of those that do can be of help. As of this writing, eighteen states and Washington, DC, are members of STEMx, also managed by Battelle. STEMx provides an environment for sharing and analyzing quality STEM education tools (see Battelle). The Teaching Institute for Excellence in STEM (TIES) offers support in STEM school design and instructional support (see Resources). At the very outset, partnering for resources such as these will help bolster the beginning efforts as the place to begin is found.

● THE SYSTEMWIDE ADVANTAGE

The challenges that exist for elementary schools are different from the ones that will exist in secondary schools. But changing what happens in one or the other, without planning for a continuous systemic shift with engaged leadership, has the potential to make the effort an exhaustive waste of time and one in which the children do not receive the full intended benefits. As children move from one level to the next, their ease of transition is related to continuity. However, if implementation can begin only on one level or the other, it is still worth beginning. **The warning is that STEM works for children at all levels, so to limit it to one level misses the advantages it affords overall.**

A STEM elementary program that does not follow students through the upper grades develops skills and expectations that are left to wither in a non-STEM high school. Having just a secondary program means students will not come with the skill set for success, and that, along with content, needs to be developed. It also allows for those students who would have become engaged learners earlier if STEM were the model to be lost by the time they enter high school. Nevertheless, most districts begin at one point with the most ready school. If that is the process, preparing the rest of the system and community should be part of the overall strategic plan.

East Syracuse Minoa Central School District

The East Syracuse Minoa Central School District (ESM), a district of 3,500 students, is located in central New York State. The district partnered with

Siemens in 2009 as part of the corporation's work on an Energy Performance Contract for the district. Siemens is an international company with business focus in electrification, automation, and digitalization. Together, Siemens and the district applied for a New York State Energy Research and Development (NYSERDA) grant that was awarded to the district. The grant provided a kiosk in the lobby of the high school that allowed students to learn more about the sources of energy in the district, including solar energy. This began a dialogue and relationship that led to what they have come to refer to as a "partnership for learning." Since that time Siemens has been a partner in ESM's development of STEM/STEAM/STREAM models for learning and the shift to transdisciplinary, project-based learning. They also partnered with the Teaching Institute for Excellence in STEM education (TIES), an organization that helps with STEM school design, STEM curriculum, and STEM instructional support to schools, districts, states, and the federal government to investigate STEM possibility.

Their work together led Superintendent DeSiato and her team to visit two premier STEM high schools in the country, one in Columbus, Ohio, and the other in Austin, Texas. Cross-organizational teams of science, math, and technology teachers joined principals and district-level leaders for the site visits. Building-level and cross-organizational teams from the middle school and the high school participated. They witnessed other learning models and other strategies that were being implemented. They met with high school juniors and seniors and asked questions. One question was, "Is there something that could have been done in your educational preparedness up to this point, which could have improved your experience in high school or preparing now for college or career, and what would it have been?" At both sites, students responded that they wished they had more of an awareness of career types and more of an understanding of the type of learning that would lend itself to the things they were now experiencing in high school. If they had possessed those two things, they indicated, they would have made better choices in their selection of courses as they progressed from ninth to tenth and then into their junior and senior years.

With that piece of data in hand, before they even got on the plane to head back home, the middle school team began thinking about what a middle school STEM program would look like. When they returned, the middle school and high school teams began to observe and acknowledge how siloed education had become. They began talking about shifting to interdisciplinary or transdisciplinary learning environments and considered how lesson design would change. They understood the need to shift from "I need you to know these particular elements or information in this particular discrete subject area, and I need to test how well you know it" to allowing students to gain knowledge through teaching and learning by having them apply it to a problem-solving methodology within the lesson design. They wanted students to take more responsibility for the application and demonstration of what they know. These planning visits by the cross-organizational teams and the ensuing conversations were essential to the choice by the district to move into a STEM shift.

REFERENCES

Collins, J. (2001). *Good to great: Why some companies make the leap . . . and others don't.* New York, NY: HarperCollins.

Drew, D. E. (2011). *STEM the tide: Reforming science, technology, engineering, and math education in America.* Baltimore, MD: Johns Hopkins University Press.

Luke, H. M. (2000). *Such stuff as dreams are made of.* New York, NY: Parabola Books.

Meadows, D. H. (2008). *Thinking in systems: A primer.* White River Junction, VT: Chelsea Green Publishing.

Scharmer, C. O. (2007). *Theory U: Learning from the future as it emerges.* Cambridge, MA: Society for Organizational Learning.

RESOURCES

 Access live links at **http://bit.ly/TheSTEMShift.**

Battelle. *STEM Education: Growing Tomorrow's Innovators in Science, Technology, Engineering and Math*: http://www.battelle.org/our-work/stem-education

Teaching Institute for Excellence in STEM (TIES): http://www.tiesteach.org

8 Planning the Shift

There is no good reason why we should fear the future, but there is every reason why we should face it seriously, neither hiding from ourselves the gravity of the problems before us nor fearing to approach those problems with the unbending, unflinching purpose to solve them aright.

—Theodore Roosevelt

S TEM can be an organizational shift, not simply a program. It can unleash the capacity dormant within a school, a district, its community, and the worldwide community if a district chooses to allow and encourage it. It dismantles silos and breaks apart impenetrable walls. It opens schools to new professional conversations and collaborations within the school and with parents and professionals within the local and global community.

If a leader and the local community hold a vision that calls for a Stage I entry, a districtwide planning effort may not be needed. For those who want to a Stage II or III shift, the structure and support of a well-designed strategic planning process are highly recommended. Most planning processes begin with the leader, and a planning team is constructed so that a vision or mission emerges. STEM shift planning may begin differently.

> A STEM shift dismantles subject silos, breaks impenetrable walls, and invites new professional conversations and collaborations locally and globally.

With full transparency, a leader who has become convinced that STEM holds the potential desired opens the conversation in an exploratory way to determine collegial interest and support. If these conversations are positive, the leader will reveal the vision as part of the invitation to join a planning team, which will serve as a STEM-shift design team.

In East Syracuse Minoa, the strategic plan articulated the district's vision to become an exemplary 21st century learning community whose graduates are prepared to excel in a complex, interconnected, changing world. Their overarching vision, mission, and goals informed their strategic action steps and plans.

At the beginning of the strategic planning process, and throughout, it is critical to remember that a STEM shift can be a program (Stage I), multiple classrooms (Stage II), or a systemic shift (Stage III) in the teaching and learning processes and in school structures and practices. Physical proximity to resources like colleges, universities, high-tech manufacturing, and health-care facilities or even talents and strengths of the faculty can become less of a limitation through technology utilization. In the strategic planning process, perceived challenges are all placed on the table but never cause the elimination of an unexplored possibility. Challenges can alter the path but need not be permitted to become insurmountable barriers. Strategic thinking actualizes the plan, but unfettered dreaming launches it initially.

East Syracuse Minoa Central School District Strategic Plan

Link also available at http://bit.ly/ TheSTEMShift

> Strategic thinking actualizes the plan but unfettered dreaming launches it.

Planning for this shift holds many possibilities if done well. However, in too many cases, strategies and interventions designed to improve schools and classrooms fail to make any real, sustainable difference to learners and learning outcomes. Although contexts inevitably vary, there are four core reasons why there has been so much innovation and so little improvement. (Harris, 2013, p. 109)

Harris cites the four core reasons as (1) reform tending to be top-down; (2) too much focus on external accountability; (3) interventions from policymakers; and (4) shift being too small, not systemic (p. 110).

- Is the STEM shift being locally chosen?

- Is external accountability driving the plan and limiting the school or district's thinking and creativity?

- Is the plan responsive to policy or to the vision of the school or district?

- Does the plan call for a real shift in the learning environment or is it just another tinkering with the existing system?

● PROCESS SELECTION

A strategic plan involves the selection of a planning process. There are many private providers of strategic planning services. Some professional organizations and school board associations also offer these services to their members. Researching the experience and cost of providers will guide the decision to select a provider who facilitates effectively and inclusively and can think disruptively as well as incrementally. You do not want a facilitator to derail the process by inserting voices of skepticism. The facilitator may be a person from the school or community, but if that is the district choice, it must be someone who has great credibility and who can facilitate without judgment. Whoever conducts the planning process must be forthright about the time commitment for planning well and the associated costs.

● THE PLANNING TEAM

The planning team will be invited to generate the plan, to guide its implementation, and to assess its success. The twenty- to thirty-member team comprises both internal and external constituencies. The internal school community representatives will include faculty, staff, school and district leaders, board of education members (not the whole board), parents, and students. External representatives can come from the fields of health care, business, higher education, the media, faith communities, and other nonprofit organizations. Transparency demands that members of the planning team understand that their work will be focused on a STEM shift. **Wise leaders will be sure to include proponents of the shift as well as skeptics.**

The planning team organized by East Syracuse Minoa Superintendent Donna DeSiato included the district's executive director of curriculum, instruction, and assessment; director of teaching and learning; board of education members; a deputy superintendent and other executive cabinet members; principals, directors, coordinators, teacher leaders, business partners, and community advocates and leaders; parent leaders and student leaders. The interest initially arose from school leaders who were looking at the research and discussing a local response, but it was not launched as a top-down initiative. The process was collaborative and inclusive, engaging the community while very clearly focusing on research and what they understood it was demanding as a response. This established why they needed and wanted to make a shift. It became clear that the rapidly changing world, over the past few decades, required a change in the structure of schools. They examined what it would take to make the shift. They understood the vast amount of work that would be needed to really retool, reequip, and build capacity within the system.

Prior to the first meeting, the STEM-shift planning team will be given the data gathered by the leader(s) in the visioning process as well as other student and district data to inform their deliberations. Guidelines for the planning teamwork are established with the facilitator and often include consensus rather than majority voting. In the end, each member can, and will, present the plan with a consistent meaning.

With a trained facilitator, a commitment to time, ready access to data, and the entry work completed, a STEM strategic plan can be generated within six months. Adding action steps, determining priorities, and budgeting for the plan will be accomplished over the following months—but at that point everyone knows the direction in which the district is moving and readiness for the shift is under way.

● THE SYSTEM'S PLAN

The strategic plan should be designed with the entire system in mind. The specific beginning point will be determined within the plan. This supports the STEM-shift capacity more than having the beginning point predetermined by the leader. When completely finalized, the plan will answer both the conceptual and the operational questions. The reasons the district wants to make a STEM shift are made clear and are communicated well and often. The resulting plan is simple, clear in its language and short—no more than a few pages. Words are chosen intentionally. The language used in the development of the strategic plan sets the shift in motion. Language that is restrictive will define boundaries and focus. Language that is inviting and invigorating will carry people forward.

The plan will include the district's mission in regard to the STEM shift, the beliefs that support the STEM decision, objectives, and strategies as well as action plans. Eventually, all this will be adopted by the board of education and broadcast widely throughout the community. The plan becomes the basis for resource allocation.

The plan will be revisited annually by the planning team to measure progress and to readjust as required. Annually, the leadership team and board reassert the STEM choice, examine data, approve the next steps, identify the human and fiscal resources and timelines, seek invested partners, and ensure that sustainability is everyone's responsibility. A link to a comprehensive STEM Immersion Guide can be found in Resources, and an annual checklist for planning can be found in the Appendix.

Granite School District

Another example of how a district created the environment in which a shift of this magnitude can be supported and nourished exists in the Granite School District in Utah. In their five-year framework, they cite a dedication to offering a very common set of objectives: improved graduation rate, increased access to technology, support for professional development for teachers and administrators, new student management system, a focus on curriculum and assessment, updated safety procedures, and added blended learning opportunities. Only these few lines are given to STEM in the entire six-page document:

- Staff and open new STEM elementary school.

- Consider creating district Information Systems redundancy site at STEM secondary school.

- Extend STEM experience to 7–12 secondary school.

Granite School District did not begin its STEM journey to follow a fad or trend. Rather, it has a vision for the advantages of integrated, problem-based learning opportunities with 21st century technology, and a focus on learning science and math as part of an overall plan. Work on the elementary level began with the building and staffing of the district's first STEM elementary school, which included both neighborhood students and those wishing to attend from outside the school's catchment area. Tyler Howe, the principal of the Neil Armstrong Academy, reported,

> This (Neil Armstrong Academy) was the first elementary school in the district dedicated to STEM specifically built for that purpose. But there exist other programs, such as McCarter Granite Technical Institute, in which all of our high school students have an opportunity to participate. They can hop on a bus in the morning or afternoon and go to a central location in the district where they are engaged not only in learning a curriculum, but they learn it hand in hand with the professionals in that industry. (personal communication, February 28, 2014)

Granite
School
District
Five-Year
Framework

Link also available at http://bit.ly/ TheSTEMShift

The plan called for creating the elementary school, followed by the upcoming opening of a STEM middle school, and then a STEM high school in 2017–2018. The district has carefully designed phasing in a K–12 STEM shift. The district is determined to provide the skills and learning opportunities for higher

grade levels and more students in the district in order for more students to be better prepared as they enter a STEM high school.

Students who may experience this manner of learning in their early years, who are placed in a century-old model in their secondary years will, of course, continue to learn. But it seems counterintuitive to innovate in the early years and return to the more traditional teacher-focused, silo-based subjects for the latter portion of their school years, especially since they will be entering colleges or workplaces that are 21st century skill settings.

In the meantime Tyler Howe and his faculty wonder what it will be like for those early graduates of the Neil Armstrong Academy to experience more traditional middle school experiences after having such an integrated, STEM-focused learning experience. While waiting for the STEM middle school to be built and opened, they are watching, hoping that what their young children gained will serve them well in a very different setting.

● MAINTAINING SHIFT MOMENTUM

A serious danger lies in a poorly planned implementation of any kind. STEM, like anything else, can be a pendulum that will lose energy and fall backward if intervening forces prevail. It seems appropriate to use a scientific explanation here. Newton's first law of motion states an object in motion stays in motion and an object at rest stays at rest, unless acted upon by an outside force. So, if the vision for the plan is based solely on preparing students for the 21st century workplace, although a laudable goal, it may be too narrow for long-term energy and momentum to be sustained. Outside forces can intervene and disrupt progress. The shift must be based on what we now know about how students learn as they make meaning of the world.

● WORKING THE PLAN

Farmers know that preparing the fields and planting crops do not necessarily result in an abundant harvest. Real work—and yes, some support from nature—is required to maximize crop yield. The same is true of the shift plan. Specific attention to the sustainability of the implanted steps, the potential roadblocks that may occur, the process to assess progress, and the type, manner, and frequency of ongoing professional development are all key drivers of the success.

Things to Remember

No matter the name chosen, *STEM, STEAM, iTEAMS*, value lies in the engaging, empowering nature of investigation, collaboration, integration in the learning process, and increased accessibility to authentic professional, real-life experiences with the field. Melissa Delaney writes for *EdTech Magazine* and offers advice culled from Dr. Jo Green-Rucker, the DeSoto (Texas) Independent School District's Assistant Superintendent for Curriculum, and others. Here are modified bits of their advice that should become part of the ongoing process involved a STEM shift:

- **Overcommunicate.** Make sure everyone in the community knows and understands the differences in the way children will be learning and that schools and days will be structured. Prepare students, parents, and teachers for the role of STEM field professionals in the K–12 learning environment.

- **Offer ongoing professional development.** Everyone will need leadership, support, new ideas, and coaching over time.

- **Give it time.** This is a process that takes time to learn, reflect, revise, and try again. Be patient and persistent.

- **Don't set it in stone.** This is a dynamic process by its very nature. It cannot be allowed to entrench and solidify and, in a few years, become as inflexibly and as structurally solid as its predecessor system. It must be kept alive and, therefore, growing and changing.

- **Make collaboration a priority.** Time must be redesigned to allow for planned time that faculty can use regularly to work together.

- **Create a culture of experimentation.** Encourage the sharing of successes and missteps and learn from both. Invite teachers to think innovatively.

- **Starting small is OK.** There are many people and places in your school ready to make the shift. A good strategic plan can allow for mini starting places to take hold throughout the system, like generative popcorn. Engage everyone in the process, but don't expect the first steps to be the same as the next. And don't expect everyone's first steps to reach the same level of success (see Resources).

Sustainability

Teachers tend to stay in a school or district for many more years than their leaders. The tendency for leaders to move from role to role and district to district often endangers sustainable change. Since the implementation of this shift requires a multiyear process, it is important to establish sustainability in ways that transcend the original leader or leaders. The choice to become a STEM-shifted system must not be top-down if its life is to endure beyond those who are its creators. Roots must be deep within the system. Unless a plan for sustainability is embedded from the initial stages, the entire shift is endangered, exhausting the participants, and atrophying from lack of nurture.

> The choice to become a STEM-shifted system must not be top-down if its life is to endure beyond those who are its creators. Roots must be deep within the system.

Anticipate Barriers

As the action plans are unfolding and the shift begins to take form, anticipating possible barriers is a leadership essential. Exploring the possible responses to those barriers helps protect those on the journey. **When navigating white water, it is good to have anticipated where the waters will be the roughest.** Even at this very early stage of planning, some of the barriers or challenges are known, and others can be predicted. Wise leaders plan for how to either avoid them or plan what to do if they arise.

Professor Peter M. Gollwitzer (1999) distinguishes between goal intentions and implementation intentions. He defines goal intentions as identifying an end point or outcome. The process of stating a goal helps commit you to the goal. Once the goal is stated, one feels an obligation, a commitment to achieve it. Gollwitzer further defines implementation intentions as ancillary to goal intentions and suggests we "specify the when, where, and how of responses leading to goal attainment." This requires the declaration of a set of circumstances that could possibly interfere with the goal attainment. Informed questions like "What will we do if influential voices in our community question our goals and intentions?" will offer the opportunity to plan for responses should these identified interfering behaviors to the goal attainment occur. Doing this in advance, rather than waiting and being blindsided, and having to plan responses on the fly, helps support the progress of the shift (see Gollwitzer, 1999).

There is the big shift to become a learning community that engages in problem- and project-based learning using STEM as a basis. And then there are those steps along the way: combining science and math classes; budgeting more money for professional development; creating relationships with businesses, organizations, and higher education; changing the schedule; or holding community meetings to engage more people in the process. For both the big goal and the smaller ones, identifying the potential problems and articulating planned behavior in response are essential. It can be as simple as a set of charts. Figure 8.1 is an example of how to begin.

Figure 8.1 Opposition Anticipation Plan

Goal	
If (Perceived or intuited opposition or barriers)	**Then** (Behaviors or steps to turn oppositions or barriers into support)

Study Others

In their book, *50 Myths & Lies That Threaten America's Public Schools,* David C. Berliner, Gene V. Glass, and Associates (2014) describe myth 27 as follows: "If a program works well in one school or district, it should be imported and expected to work well elsewhere" (p. 129). **It is not possible to lift a program from one place and lower it into another and expect the same result.** The variables that exist are extraordinary. School culture, socioeconomic levels, competing initiatives, relationship with the community, core values, funding, staffing, and so on over an exhaustingly long list are variables that require decisions be made locally. Schools are complex entities. They cannot duplicate programs and expect the same results as were achieved in other places (Berliner, Glass, & Associates, 2014). However, the travel to see models in the beginning stages of their work was reported as having great value by the schools researched for this book. All reported speaking to those who were steps ahead and seeing those schools in action. Doing so helped inform clarifying questions, offered answers, and gave rise to yet unthought-of ideas—and reinforced the decision to shift.

Teachers as Partners

The systemic shift can be overwhelming. Planning well helps to capture the facets of the change and track of movement, successes, stuck places, and things that still need to change. Another example of a comprehensive 21st century strategic plan can be found on the website of Goochland County Public Schools. The vision as a 21st century school applies to the teachers in the classroom through the ranks of the leadership to the board of education. They know where they want to go and how they will go there.

In order to make a total systemic change, they included a change in the way they addressed career and technical education (CTE) as a facet of their plan. CTE, while more applied than traditional academic programs in most high schools, still forces credit-bearing courses to be taught in the traditional subject silos. Superintendent James F. Lane explains,

Goochland County Schools Strategic Plan

Link also available at http://bit.ly/ TheSTEMShift

> We're trying to break that down and understanding the cross-curricular connections between how CTE relates to what's going on in the science and mathematics classes. The move toward robotics and engineering and what I would call the modern Career and Technical Education curricula has been really impressive. We went from a high school that really had little to no modern Career and Technical education to now really offering the cutting edge of what's out there for our students . . . G21 is a pre-K through 12 initiative. Our CTE students are able to go back into the English classroom, or back into the foreign language classroom or math classroom and, when they do their G21 projects (as required by the G21 plan), they are able to bring a level of expertise back to those classes because of the experiences they have had in their CTE classes. Our strategic plan is focused on getting ahead of the curve and making sure that we have engaging experiences going on in every classroom, every day. (personal communication, January 30, 2014)

In addition, they developed a plan to support the Strategic Plan. It is called G21 and requires that teachers begin by completing one project each year.

Goochland County Schools G21

Link also available at http://bit.ly/ TheSTEMShift

The project planning process begins every September. Each teacher works with the instructional technologist to design a plan for engaging students with rich content through project-based learning opportunities. Dr. Lane explains, "It's in our plans. Whether it's through new hiring or through the observation and evaluation process, we all know that to be a teacher in Goochland means that you're going to think and teach this way over the long haul." In order to make this value for teaching sustainable, they are planning to create a certification in their county, where a teacher will be G21 certified in order to be a master teacher. And to receive the most prestigious positions, such as department chair and teacher leadership positions, there will be a process that will reveal a history of excellence in this area.

> It is the live encounter among teacher, STEM professional, and students that makes a STEM shift fresh and creative.

The teachers in any system influence the success of any plan. It is the live encounter among teacher, STEM professional, and students that make a STEM-shift fresh and creative. Teachers will both follow and lead the strategic plan if they can see it come alive in the students. Even the skeptics will be open to considering its potential. Leaders need to articulate the purpose and the plan and implement it with integrity throughout the system. The attention to supporting the teachers as they implement more of this shift and as they have the opportunity to renew and refresh lessons is evidence of the district leadership's commitment to a strategic, systemic shift. Goochland's investment in the teachers, consistent reinforcement of their articulated values, and unwavering attention to the district's strategic plan and the G21 document, as well as open and constant communication with the community, all strengthen the architecture of their vision and plan, making it strategic and enduring.

Assessment

Nothing will be successful in education these days if it does not have a core element of assessment within it. Assessment for accountability has become one of the most polarizing aspects of our work. The desire for data to support decisions and guide decision making toward growth of all students has caused us to measure everything we do. The dual pressures of attention and demand from policymakers, reporters, parents, community members, and businesses have produced an environment in which only test scores matter. We have become accountable for demonstrating academic progress and achievement through assessments. In many cases, those are standardized tests. Accountability has taken on a negative connotation within the system. Education is, however, publicly funded and socially and economically essential. It is reasonable, then, to expect that we must demonstrate our effectiveness and success through some form of data.

Even in this data-hungry climate, we can make our own choices. We can include and publicize good local assessments while meeting mandates. In Goochland, Virginia, the school district leaders created a "balanced assessment" system. It still includes achievement measures required by federal law and the growth measures required by local policies but allows for recognition that these existing growth measures are not the measures for this type of learning. Under the leadership of Dr. Stephen A. Geyer, the assistant superintendent, they are piloting their first run of performance-based assessment for students. Superintendent Lane reports, "Ultimately, when it comes to evaluating this model, we want the performance based assessments to be the major focus of how we assess whether students are doing well. We want the growth assessment to be a by-product of teaching the right way."

In East Syracuse Minoa Central School District, along with a sweeping systemwide STEM shift, Dr. DeSiato reports that they watch with assurance that student engagement is one of the outcomes of this shift. The educators in ESM continuously report that there is a tremendous increase in student engagement. Parents talk about the fact that their son or daughter has a renewed love of learning and therefore can't wait to get to school to solve the problem that they are working on. They report this is true even with students who traditionally may have been really challenged. What they are finding is that perpetuating the models that weren't working was not a solution. Having a model that is interdisciplinary, transdisciplinary, and project based, and that taps into other types of skills for students, along with being very motivational and engaging, is revealing a tremendous amount of success, including students who in previous school years were not successful.

From their own measures, they are clearly seeing the evidence of improvement. Dr. DeSiato has clear evidence that students' confidence in their learning and their ability to communicate with their peers and adults are, without question, both recognized and observed by the educators within the system, parents, business partners, and by those who actually get to observe the students in action. "Hands down people always comment to us about that." In addition, Dr. DeSiato reports that even in the class held at the high school before the school day begins, no student is late and no student is absent. Students are engaged in their learning. Student engagement is the essential ingredient for achievement. They are certain results in even the required standardized tests will rise as students' interest and attention to their learning continues to grow.

We give assessments all of the time and use their results to drive some of our decisions. But most educators are not well trained as assessment developers. In a system in which students are asked to learn and problem-solve alone and with their peers, will traditional assessment practices measure achievement? How will we be able to track progress in the diversity of skills, abilities, and information as students work together and alone to answer questions, solve problems, and build new ideas? If we weren't experts in the first paradigm, how will we be in the next?

The strategic plan must hold a space for learning best practices in assessment, particularly in this new paradigm for learning. If we leave assessment out of the plan, we will find ourselves in the middle of an entirely new paradigm for learning without a way to answer the call for the data needed for accountability for students, teachers, and schools. It will be an indefensible position in the current environment. **As we design schools for the future, the knowledge and capacity for designing appropriate assessments must grow along with it.** "STEM PBL [project-based learning] requires a whole new perspective on what assessment means" (Capraro, Capraro, & Morgan, 2013, p. 111).

In addition to the limits that exist in our own expertise with assessment, students are accustomed to assessment as the route to a grade. By high school, the action of using assessment to quantify knowledge attainment has become an expectation of students, and certainly their parents. In this new teaching and learning paradigm, formative assessment is as important, if not more important, than the summative assessment students and parents have come to understand and respect. Although most assessments can be used for formative or summative purposes, it is the formative assessment that informs students of their progress and development both individually and as a group. It is the formative assessment process that allows students to receive support and direction from their teachers. **Skills develop, understanding grows, and knowledge blossoms, all before the learning event is over.**

Authentic assessment "is the most complicated assessment method compared to other formative and summative schemes . . . there is a consensus among educators that authentic assessment tasks should focus on the knowledge products, which make the assessment relevant to the learner through real-world applications" (Capraro et al., 2013, p. 111). The authentic nature of STEM learning opportunities should be paired with authentic assessments. Teachers and leaders need to learn a new process for assessment development and use. Parents and students need a planned, gradual unfolding of new understanding about the difference in the manner in which assessment will take place. This must be part of the strategic plan.

REFERENCES

Berliner, D. C., Glass, G. V., & Associates. (2014). *50 myths & lies that threaten America's public schools: The real crisis in education.* New York, NY: Teachers College Press.

Capraro, R. M., Capraro, M. M., & Morgan, J. R. (Eds.). (2013). *STEM project-based learning: An integrated science, technology, engineering, and mathematics (STEM) approach* (2nd ed.). Boston, MA: Sense Publishers.

Harris, A. (2013). Building the collective capacity for system change. In H. J. Malone (Ed.), *Leading educational change: Global issues, challenges, and lessons on whole-system reform* (pp. 109–113). New York, NY: Teachers College Press.

Malone, H. J. (Ed.). (2013). *Leading educational change: Global issues, challenges, and lessons on whole-system reform.* New York, NY: Teachers College Press.

RESOURCES

 Access live links at **http://bit.ly/TheSTEMShift.**

Arizona Stem Network. (2013). *The STEM Immersion Guide for Schools and Districts*: http://stemguide.sfaz.org/wp-content/uploads/2015/01/SFAz_STEM_ImmersionGuide1214.pdf

Delaney, M. (2014). *7 Guidelines for Building a STEAM Program*: http://www.edtechmagazine.com/k12/article/2014/04/7-guidelines-building-steam-program

East Syracuse Minoa Central School District. *Strategic Plan*: http://www.esmschools.org/district.cfm?subpage=24324 **(QR code on page 75)**

Gollwitzer, P. (1999). Implementation intentions: Strong effects of simple plans. *American Psychologist, 54*, 493–503 (research on preparing for obstacles before they arise): http://www.psychology.nottingham.ac.uk/staff/msh/mh_teaching_site_files/teaching_pdfs/C83SPE_lecture3/Gollwitzer%20(1999).pdf

Goochland County Schools. *G21: A Framework for Developing Twenty-First Century Skills and Deeper Learning Experiences*: http://www.glnd.k12.va.us/index/resource/g21/ **(QR code on page 83)**

Goochland County Schools. *Strategic Plan*: http://www.glnd.k12.va.us/index/schoolboard/plan **(QR code on page 83)**

Granite School District Five-Year Framework: http://www.graniteschools.org/teachinglearning/wp-content/uploads/sites/13/2014/11/GSD-5-Year-Plan.pdf **(QR code on page 78)**

9 STEM Curriculum Shifting

*It was an initiation into the love of learning, of learning how to learn . . . as a matter of **interdisciplinary** cognition—that is, learning to know something by its relation to something else.*

—Leonard Bernstein

Problem-based and project-based learning (PBL) opportunities develop students' conceptual knowledge. Well designed, they are fundamentally interdisciplinary and collaborative (Capraro, Capraro, & Morgan, 2013, p. 51). Schools making the STEM shift develop dynamic planning environments in which teachers continually grow and, consequently, offer increasingly rich, integrated, empowering, skill- and content-based learning opportunities for children.

> Schools making the STEM shift develop dynamic planning environments in which teachers continually grow and, consequently, offer increasingly rich, integrated, empowering, skill- and content-based learning opportunities for children.

In 1992, Dr. Eli Eisenberg began working with then President Nelson Mandela on the Science and Technology Education Project (STEP) for South Africa to address the educational needs of the population and the problem of unemployment in that country. Why take a poor, struggling, newly organized country and think about changing schools to be organized around science and technology? Dr. Eisenberg explained. At that time in South Africa, children lacked the resources and experiences that would help them develop as strong academic students. However, students in South Africa were particularly strong in art and music and drama. Integrating those subjects with mathematics, science, and technology was the place to begin their new educational system. This organization of STEM-based integration of subjects, based on the strengths of the students, would then provide "authorization, strengths, and self-esteem." For Dr. Eisenberg and those with whom he works worldwide, STEM is not just a program. It is a real systemic change in the way teaching and learning take place.

According to Dr. Eisenberg, when you walk into a STEM classroom, you may not see textbooks or the traditional way of learning starting from a book. You see groups of students questioning an issue, debating about an issue. The problem or issue is introduced in a general way. It is based on a real-world situation in very general words. **The books don't hold the answers, and the learning activity is "experienced" in STEM education.**

Dr. Eisenberg believes innovation is something one experiences. The student has to do the work, think much more deeply, without being given the answers. Questions cannot require an answer of "yes" or "no." The questions have to require answers that may involve a defense of the answer with evidence. Dr. Eisenberg notes that students should learn how to select the appropriate information and discard what is not relevant. The students have to be able to know and communicate how they will proceed with a project, or how the process being used is a good way to answer the problem they are facing. Dr. Eisenberg asserts that the teacher's role should be leading the course and that the teacher should be an expert in knowing how to ask the questions and what types of questions should be asked. He refers to the **Socratic way of learning in which the teacher becomes a master questioner,** which he acknowledges is a highly disciplined ability. A Socratic questioner must be able to keep the discussion focused and intellectually responsible. Essential to the process, the questions asked must be probing and open ended. At selected times, a decision must be made as to whether the teacher or the students summarize what has taken place and what has yet to be resolved. And the goal of engaging as many students in the discussion as possible is met (Paul & Elder, 2006). STEM is first about how the learning takes place and second about science, technology, engineering, and math.

Dr. Rosemary Millham, former high school teacher, NASA project manager, published author, and currently associate professor of secondary education at SUNY New Paltz, offers her perspective on the potential for STEM education. A key challenge within STEM education is that "there really isn't any delineation in the subject matter." STEM is an interdisciplinary practice. Dr. Millham states,

If you build STEM education on four isolated topics, you are doing a disservice to STEM. It is science, technology, engineering, and mathematics, and one cannot exist without the other. You cannot teach science without history. You cannot teach science without art, mathematics or literature . . . The reason STEM hasn't had a successful shift in our schools is our failure in making it a systemic shift and simply offering more classes, or clubs, or activities. How prepared are teachers to incorporate science, technology, engineering and mathematics and connect them? These subjects are disconnected over the P–12 years. In high school they are disconnected by schedule and teacher certification rules. If we are going to make a difference it has to be at every grade level. (personal communication, February 24, 2014)

A key curricular facet of a true STEM shift is the presentation of planned opportunities for students to apply learned information and concepts to new problems and questions. Application of knowledge is an essential part of the curriculum and includes ever-increasing levels of difficulty and a variety of opportunities for application alone and with others. Curriculum implementation is a continuous process where reflection and adjustments are ongoing, as illustrated in Figure 9.1.

Figure 9.1 Curriculum Implementation Cycle

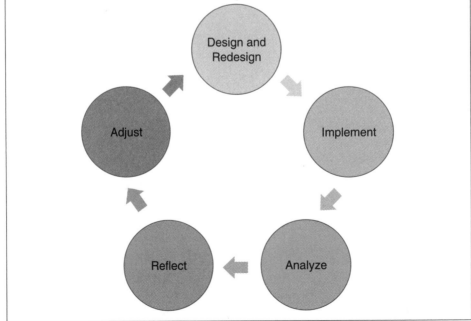

● ELEMENTARY

Integrated transsubject lessons are already taught in many elementary classrooms. The math, reading, writing, science, social studies, art, music, and technology used in lessons can and should be aligned with the curriculum and standards for those subjects. Elementary teachers are natural integrators. Therefore, there are many elementary teachers who have managed to avoid separating their teaching into subject bites, and have, for example, taught reading and math in a science lesson, or have developed project-based themes, perhaps more than we imply here. For all who have done that, a step toward the STEM shift is already completed. These are pockets of readiness.

Once we are prepared to begin developing this type of interdisciplinary learning, the starting place is a very local decision. The integration of subjects provides the opportunities for students to understand, from early on, the relationship these "subjects" have with each other in the real world. Or, to flip this perspective, it is not until school that children know subjects are separate. They experience the world as a unified place before school teaches them otherwise.

> Children experience the world as a unified place before school teaches them otherwise.

Engineering has not been traditionally a delineated part of the K–12 curriculum. However, as STEM has become a point of focus, schools have begun to develop engineering courses, particularly in the high school. More recently, elementary options are also appearing. The decades old Project Lead the Way (PLTW; see Resources), a national leader in providing training and curriculum in STEM education, recently received funding from the Louis Calder Foundation to develop a PLTW Launch Program for students in Grades K–5. PLTW CEO Dr. Vince Bertram's opinion is that engaging students in math and science in their early years helps them develop a "life-long love of these subjects" (see Project Lead the Way, 2014). Children engaged in the STEM shift, from their early years in kindergarten, are more likely to be better prepared for more rigorous project- and problem-based learning in their secondary years. That is a good thing.

Randolph Elementary School, Goochland County Public Schools

Beth Ferguson is a fourth-grade teacher at Randolph Elementary. In a unit on plant studies, she created the environment to engage her students as collaborative investigators. They learned through a dynamic ten-lesson unit in which they actively worked with the Randolph Elementary School

Courtyard Team. The team was made up of landscapers, teachers, community members, the 4-H, and a master gardener from J. Sargeant Reynolds Community College who came to the school to help ignite the interest of the entire fourth grade. The capstone to the unit was designing a plan for vegetation that would thrive in the courtyard of their school for years to come. Students had to research plant life cycles and weather. They made observations, performed dissections, developed a garden sustainability plan, considered intervening variables, conducted research, and collaborated with professionals in the field. Students had to read research that was written above their grade level, interpret and make sense of it, write about it, and prepare a presentation. They learned and were evaluated on information from their social studies, science, math, communications, English language arts, technology, and art curricula. Students were team members and independent researchers. Throughout the unit, students were engaged in planned interdisciplinary inquiry in which the teacher served as a guide and was able to intervene with minilessons as the need arose. This is the design for all units of study in her classroom.

Beth Ferguson commented on her district's progress in their implementation of STEM:

> I think what we are starting to understand is this is a whole other way of teaching. It's a lot more efficient. Children gain a lot more knowledge. There's a whole lot more inquiry. When there's inquiry they're going to continue wanting to learn and retain more information. We feel pretty comfortable in teaching cross-curricular subjects, but we now have to really think about how we can become better teachers by asking better questions and pulling more from our students. What we have learned to do in our science teaching we incorporate into the other subject areas as well . . . and it all gets incorporated together. (personal communication, January 30, 2014)

Hattie Cotton STEM Magnet Elementary School, Metro Nashville Public Schools

The Hattie Cotton STEM Magnet Elementary School in Metro Nashville, Tennessee, shifted from being a traditional elementary to a STEM Magnet in the 2011–2012 school year. In January of that school year, Dr. Vicki Metzgar, now Director of the Middle Tennessee STEM Innovation Hub, was the Metro Nashville Public Schools' STEM Director. She developed a team of instructional designers to lead and support the STEM shift. The instructional designers assigned to Hattie Cotton Elementary School worked with the building leader and faculty on the development of integrated, problem-, and project-based learning opportunities. This involved the redesign of curriculum, training, technology support, and securing supplies and equipment. The collaborative leadership of the principal and the instructional designers created an environment in which **teachers ventured out of rooms to become curious, innovative risk takers**.

Every nine weeks teachers taught a different unit. The themes were Friends Far and Near; Circle of Life; Wind, Water and the World; and Investigations of Transformations. The coaching and modeling, thematic planning, collaboration and integration resulted in extraordinary growth in achievement scores at the end of that first year. Hattie Cotton earned a spot on the Tennessee Department of Education's Inaugural List of Reward Schools for top growth in student achievement.

Reflecting and **redesigning** are embedded in the ongoing process of revising the learning opportunities. Reflect, refine, and revise are parts of the ongoing curriculum development process of the STEM shift as illustrated in Figure 9.1.

● SECONDARY

STEM programs that begin in high schools have tended to be more centered on the four subjects as they are related to career fields. Even if the courses are open to all students, those who have not excelled in science, technology, engineering, and math or have not discovered their own interest in them are often not included. Too often, these will be the "special population" students discussed in Chapter 5.

The other form of secondary start-up is a STEM-focused charter school or magnet school, a school of choice. In these ways, STEM becomes an alternative place rather than a shift in an existing one. These are not bad options, but the benefits of the STEM shift are for all. Bravo to those who take on making STEM shift within the system and making it really public—for everyone.

Until students begin arriving with more developed problem-solving and mathematical and scientific skills and are better prepared to use the technologies available, flexible opportunities must exist. So even if the high school is the building chosen for the shift start-up, involving the K–8 leaders and teachers in the choices and the early professional development will encourage the implementation of problem-based, STEM-focused teaching and learning in projects before high school.

We need to do this because scholars have noted that **formal educational environments have been better at selecting talent than developing it**

(Bloom, as cited in Bransford, Brown, & Cocking, 2000, p. 5). By the time students enter high school, they have already been selected in one way or another, have developed their own beliefs about who they are and what they are capable of doing, and only with very individual intervention in the context of a strong relationship will this become malleable. So at the outset, it is important to create developmental course offerings that are integrated and invite all students to be problem solvers and independent thinkers. This will help to narrow the gap until students who are more experienced with this type of learning arrive on the high school's doorstep. Until a student has had the experience of this type of problem-based, student-centered, integrated learning over the course of those nine precious years before high school, the high school program design will have to incorporate an entry plan to introduce and prepare students to learn in this new approach. Over time, students will come prepared after a systemwide shift is complete.

The developing skills and abilities of the K–8 student population engaged in STEM learning will demand that the high school STEM design be a dynamic and changing one in order to meet the needs of the students as they arrive (see Figure 9.2).

Figure 9.2 High School Change Accelerates K-8 STEM

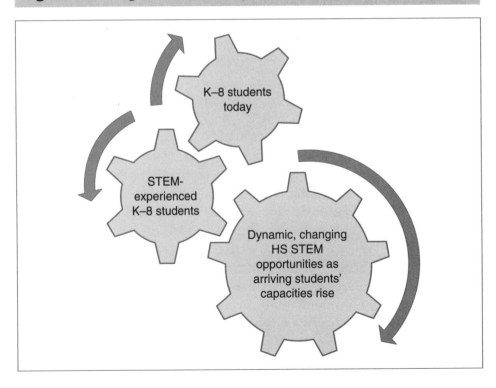

Bransford explains that

> Many people who had difficulty in school might have prospered if the new ideas about effective instructional practices had been available. Furthermore, given new instructional practices, even those who did well in traditional learning environments, might have developed skills, knowledge, and attitudes that would have significantly enhanced their achievements . . . Learning research suggests that there are new ways to introduce students to traditional subjects, such as mathematics, science, history and literature, and that these new approaches make it possible for the majority of individuals to develop a deep understanding of important subject matter. (pp. 5–6)

DuPont Hadley Middle School, Metro Nashville Public Schools

DuPont Hadley was not one of Metro Nashville's magnet schools, but it was required to move to a problem- and project-based learning design. The Buck Institute for Education (BIE) creates, gathers, and shares high-quality PBL instructional practices and products, and provides highly effective services to teachers, schools, and districts. DuPont Hadley's entire fifth-grade team attended training at the BIE, where they developed a four-week unit of study devoted to cells. Subjects included were math, English language arts (ELA), social studies, science, creative writing, technology, and band.

Each part of the lessons and assessments were aligned with the Common Core Standards. Grade-level standards for mathematics included decimals and multiplication. Grade-level standards for ELA included conducting a short research project to answer a question, clear and coherent written arguments, revising, and editing. Grade-level standards for social studies (economics) included describing the potential costs and benefits of personal choices in a market economy. Grade-level standards in science included distinguishing between the basic structures and functions of plant and animal cells, identifying major parts of plant and animal cells, and comparing and contrasting basic structure and functions of plant and animal cells.

Formative assessments included tests and quizzes, journals and learning logs, rough drafts, online tests, practice presentations, notes, and checklists. Summative assessments included written products (persuasive letter and a budget), multiple choice and short answer tests, a performance rubric, peer evaluations and self-evaluations. The manner of reflection required was journal writing, whole-class discussion, and surveys.

Teachers Pam Newman and Caroline Neuffer reported that their project culminated in an evening event with a fund-raising dinner and presentation by the students. "They could stand up and present to the community what they had learned and shared with strangers that they didn't know. That's very difficult for fifth graders. A lot of them stood up before a live audience. And we were really impressed with their performance."

These teachers saw evidence of the value of project-based learning in the quality of the writing the students produced. Reflecting on this past year and looking forward, they shared what was learned and what might change:

> The students learned the process that you go through in doing research. And, reflecting over the project, some of the teachers were saying that we gave the students voice and choice, but next time we need to give them more voice and choice, and we need to involve them more in the discovery process. We need a little bit more time so they have the time to explore and discover things on their own rather than us sort of leading them through the process.

Stratford STEM Magnet High School, Metro Nashville Public Schools

High School Principal Michael Steele (2013) clarifies the place of curriculum in a STEM-centric school:

Stratford STEM Magnet High School

Link also available at http://bit.ly/TheSTEMShift

A STEM high school is not a curriculum. A STEM high school is a way of doing business. It's a way of thinking. And we do everything we possibly can through project-based instruction. We want our kids working together. We want them working independently. We want them teaching. We want them presenting. And we want them to have a love of learning . . . Females are very scarce in engineering, and black females are almost non-existent. So, we've got to teach them early on to develop a love, or at least a strong interest, so when they get to high school, they choose those career paths. But, STEM education is a way of doing business. It's a way of thinking. It's not a way of saying hey let's do more math and science classes. But it's a way of being able to turn any part of the curriculum into a science, technology, engineering or math type of format and then being able to really come up with a way of having extremely high order questioning down to just regular high order questioning, but no low order questioning. Get kids thinking outside the box or thinking for themselves. Also what comes with that is how you're going to assess these kids. So you have to make sure your assessment process aligns with your STEM focus.

Before the move to be a STEM magnet high school, Stratford High School was engaged in a program that allowed students to spend one day a week at Vanderbilt University's Center for Science Outreach working alongside research scientists. It gave students the opportunity to work side by side with scientists but also required them to miss one day of school and be responsible each week for making up those lost classes. Now, in collaboration with the Center for Science Outreach, Stratford has a scientist in residence at the high school, every day, teaching students how to conduct and present research. These scientists coteach with the high school science teachers in a program called Interdisciplinary Science Research. This four-year course carries with it honors status.

Students can enroll in the Engineering Program or the Interdisciplinary Science and Research (ISR) Program. Both are in partnership with Vanderbilt University Center for Science Outreach. In addition to the Academy of Science and Engineering, Stratford offers an Academy of National Safety and Securities Technology. Through this Academy, students study robotic transportation, national security, and security technologies (see Metro Nashville, 2015).

Stratford STEM Magnet High School is organized into academies. The first is the Academy of Science and Engineering Pathways: Interdisciplinary Science Pathway, Engineering Pathway, and Biotechnology Pathway, and second, the Academies of Safety and Securities Technologies Pathways: National Security Technology Pathway (Criminal Justice/Forensic Science) and Computer Simulation and Game Programming Pathway.

Pathways are chosen by the students at the end of ninth grade, following a Freshman Academy experience. In addition to the standard academic requirements in English, math, science, and social studies, these pathways are chosen as electives. All of their projects become interdisciplinary and theme based.

Stratford
High School
Profile

Link also
available at
http://bit.ly/
TheSTEMShift

New Milford High School, New Milford School District

New Milford High School in New Jersey developed an academy system within its traditional high school program: The Academy of Arts and Letters, The Academy for Global Leadership, and The STEM Academy. Students entering the high school are offered the opportunity to apply for entrance into one of the academies. Admission to any of the academies is open to all students in the district. The high school requires applicants to write a statement of interest, which is to be signed by them and their parents. They are also required to write an essay explaining why enrollment in the academy of their choice is important to them. The essay includes "what attributes they possess that will contribute positively and productively to the Academy community, what they have done personally or through school that illustrates their interest, and what they expect to gain from the Academy experience." In addition, applicants are required to obtain a letter of reference from one of their current teachers endorsing their application.

This is from their brochure:

> The STEM Academy prepares students for professions in many fields including but not limited to: psychology, environmental studies, biochemistry, medicine, engineering, accounting, education, architecture, aeronautics, statistics, computer sciences, food sciences, and applied mathematics . . . The goal for STEM students is to cultivate a group of future leaders who possess the ability to think critically as a result of deep analysis. The program encourages the students to engage in self-reflection, which in turn, helps develop solid content knowledge, personal consciousness, ethical behavior and active contribution to the STEM workforce. (New Milford, 2014)

Principal Eric Sheninger shared that

> The academy model is not stand-alone. It is a hybrid, embedded approach, so basically our students are still taking their normal course work required for New Jersey graduation requirements. When a student decides that they want to be a part of the academy, they then work with their Guidance Counselors to strategically select the courses that align with their academy designation. In this case, it's STEM, but they also have Global Leadership Arts and Letters. They don't have separate areas in the building. They are not separating the students and it is not an elitist program because every student, regardless of GPA, has the right to be in the academy . . . It's all about following passions, deciding on a mini-major, taking more rigorous courses throughout their high school career. (personal communication, May 23, 2014)

New Milford also offers virtual high school courses through the Virtual High School Collaborative in which students take courses aligned with their academy. Mr. Sheninger reported that every student in the academy has experiences that include a book-study course and an independent open courseware study. STEM Academy students can access online courses at Harvard, Yale, and MIT. Students are required to complete a capstone project related to the STEM academy and in their senior year, if they choose, they can go out on a "senior service" related to their academy. In New Milford High School, as in Stratford STEM Magnet High School and East Syracuse Minoa Central High School, students are offered opportunities to go offsite to have learning experiences with an ever-growing number of partnerships the school develops with different organizations.

Two Teachers Cross Content Areas and Create New Learning

Philip Pietrangelo, a chemistry teacher working in the Diman Regional Vocational Technical High School in Fall River, Massachusetts, was concerned with the hundred-year-old model of teaching and learning. Pietrangelo had been thinking about creating real-life, interdisciplinary, project-based learning opportunities for his students.

> I felt there was so much more that could be done between the vocational and the academic side of our school. I started conversations with one of the chefs who worked in the culinary arts program. It was a state of the art department, and they had a restaurant that was open to the public 5 days a week, so they had the ability to produce a lot of food. But, on top of that, we had the ability to teach students about food in a very scientifically conducive way. We started having conversations and attended a professional development opportunity in Louisville that helped us blend our curriculum and approached administration for approval. (personal communication, October 14, 2014)

According to Pietrangelo, Diman Regional Vocational Technical High School enjoyed a long-standing reputation for preparing students for a variety careers. He explained further,

> There wasn't an after-school science club. We wanted to see if we could get kids participating in after school activities, especially for the non-sports-oriented kids. We felt like this was a trend that STEM needed to move toward, integrating content, taking a vocational curriculum and making sure that it incorporated science, and vice-versa, that science was incorporating real-life applications. We felt like there was so much benefit.

Because of schedule limitations, what was approved was a one-credit after-school club. They called it Food Science. Students learned about food in a scientific way and "got a chance to eat a lot of really cool stuff." In their freshman year, they could select their shop. Over the next four years, they would go between their shop experiences and their academic classes. It was a relatively successful schedule and situation. But Pietrangelo had a lot of students who were in the culinary arts shop, and a lot of students who were interested in culinary arts but couldn't be in that shop.

In the design of the Food Science opportunity, they asked themselves, "Where can we integrate food into my curriculum (chemistry) to make it exciting and accessible for our students?" And in their work together, Pietrangelo and his colleague realized that almost everything covered in chemistry can be applied through a food lens. An example is making ice cream—in the process of flash-freezing cream, very fine crystals are formed, giving one a very creamy ice cream. "Even the experience of working with liquid nitrogen proved so valuable for the connections they made between matter and energy and basic scientific understanding, and they got to eat ice cream." Another marriage between cooking and chemistry was the sous vide method. Cooking turducken offered an opportunity to use an enzyme to bind the turkey to the duck to the chicken and then use the sous vide method to cook it. "We were going for an experiential way. Inexpensive steak came out like filet mignon. They were learning about food with the main focus on science."

Students were required to keep a lab notebook, and they had to lay everything out in a very procedure-oriented way so that it was a science a lab but in the kitchen.

The Food Science club met for two hours once a week, offered as a one-credit class with an assigned grade. A completed final project was their final exam. It was a class/club hybrid. Students who participated in the hybrid returned to their chemistry class with increased interest in learning because what they learned in chemistry was applicable to what they could do in the kitchen.

Pietrangelo believes

> What we are doing in traditional classrooms could be so much better. This was so much fun for us and the kids . . . something our traditional classrooms are lacking, and we need to fix it. It invigorated me to pursue these types of long-term, project-based curriculum as opposed to curriculum with projects, making the curriculum thematic and experiential in nature.

One STEM Classroom's University Partnership, Montgomery County Public Schools

George Mayo (2014) is a teacher at Montgomery Blair High School in the Montgomery County Public Schools, Maryland. The school system in which he works is focused on developing students who can "think critically and creatively to solve complex problem" and whose core values, stated on their website, in part reads as

> we will encourage and support critical thinking, problem solving, active questioning, and risk taking to continuously improve; stimulate discovery by engaging students in relevant and rigorous academic, social, and emotional learning experiences; and challenge ourselves to analyze and reflect upon evidence to improve our practices. (See Montgomery County)

His motivation was a *New York Times* article written by Matteo Pericoli, an architect, teacher, illustrator, and author of *The City Out My Window: 63 Views on New York* and, most recently, *London Unfurled* (See Pericoli, 2013) In the article, Pericoli reported

> So what happens when we ask writers to try their hand at architecture? At the "Laboratory of Literary Architecture," . . . I encourage students to find—or, rather, extract—and then physically build the literary architecture of a text.

The early August article struck a chord with Mr. Mayo. A sparked interest called him to develop his own project-based, themed unit combining literature and architecture. Because it was summer, he devoted the following three weeks to developing and planning the course. The eighty-minute high school block schedule for classes allowed him to design an interdisciplinary experience with time and space for real building. Skype offered him opportunities with Pericoli, his students at Columbia University, and Dr. Boyer, professor of creative writing at Columbia University. She is the professor who worked with Pericoli having her students create structures to represent text.

Mr. Mayo reached out to another architect, a professor of architecture at Catholic University and Skyped with two graduate architectural students from the University of Maryland. In his words, "I shopped it out. I made up for my lack of knowledge by bringing people into the classroom to help us."

Without working in a STEM-shifting school and without any prompting other than his own inspiration and the *New York Times*, Mr. Mayo turned his tenth-grade English classroom into a model-building workshop, complete with materials, glue, hobby knives, and cutting boards (Mayo, 2014). The typical English classroom was gone.

Mr. Mayo described his school as having all kinds of interesting projects happening. The leadership allows for and encourages innovative teaching and learning, and it has been that way, he reports, for quite some time. Should his school decide to make a STEM shift, Mr. Mayo and his innovative colleagues will be essential partners to engage in the strategic planning and implementation of the shift. They have planted the seeds of the shift right in their classrooms.

● CLUBS AND SCIENCE FAIRS

Clubs and camps offer a low-stakes environment in which to experiment with the STEM shift. In the low-stakes environment, flexibility and creativity are more easily ignited. All environments can be considered sandboxes for experimentation and learning for both the faculty and the students. Clubs are often focused on problems and solutions.

But what is the likelihood of students who do not excel in math or science, technology, or engineering to join a STEM club or choose a STEM camp, or participate in an extracurricular science fair–type competition? Intentionality for involving the typically excluded population into these opportunities can begin to open the pathway.

> All environments can be considered sandboxes for experimentation and learning for both the faculty and the students.

There certainly are teachers who already understand the precepts of this type of learning and performing. Embedding math and architecture with the drama club and the set design seems simple and almost necessary. Interest in STEM is spawning a proliferation of science fairs across the country. These invite students to showcase projects on which they have been working with teachers and often as a member of a team. They present another opportunity to involve a STEM approach with low risk, excitement, and little opposition.

Eesha Khare Inventing the One Minute Mobile Phone Charger

Link also available at http://bit.ly/ TheSTEMShift

In January of 2014, over 400 middle and high school students from Sarasota County, Florida, schools participated in a countywide science fair. They discussed and demonstrated everything from drone trackers to water processing and fusion projects. Winners went on to the national fair sponsored by Intel International Science and Engineering Fair (ISEF) in Los Angeles, California.

The 2014 ISEF attracted 1,700 students. The top prize, a $75,000 award, went to fifteen-year-old Nathan Han of Boston for developing a machine learning software tool to study mutations of a gene linked to breast cancer. Previous award winner, California high school student Eesha Khare, is now a Harvard student. As a high school student, she invented a supercapacitor that could charge cell phones in seconds and has vast implications for the field of energy. She is also a dancer. She describes how those two paths are connected in her process.

Another Intel-sponsored event, the Science Talent Search, yielded seventeen-year-old Eric S. Chen of San Diego, California, the top award of $100,000 for his research of potential new drugs for treating influenza. The White House, too, has joined the STEM Fair wagon. This year's STEM Fair brought teams of over one hundred students from more than thirty states to the White House to present their research. The President viewed the event and discussed the projects with the future scientists, inventors, and engineers.

President Obama Speaks at the 2014 White House Science Fair

Link also available at http://bit.ly/ TheSTEMShift

CURRICULUM MAPPING

In a STEM-based school, no longer are blocks of time saved for "the multiplication unit" because those skills will be embedded as lessons included in some of the scientific investigations in which students will be engaged. Keeping track of topics, standards, skills, and progress can be challenging. Documentation with digital curriculum mapping products available on the market can be helpful in this endeavor. This is true, particularly because they can include the entire curriculum (scope and sequence), whatever standards are being addressed, and they can be cross-referenced with the lessons and units as they are entered and reviewed. However, if your school is not using a mapping product, spreadsheets, Wikispaces, or Google Drive can be places to start gathering the work as it is developed, shared, and revised along the way. It is important to have a way to keep track of the units planned, lessons taught, assessments administered, and their alignment to the standards.

Other challenges are likely to arise in which articulated changes in curriculum must be captured in order for them to be memorialized and edited, all while being shared. One such example can be found in a systemwide discovery that was made after the first year of STEM implementation in Metro Nashville. The K–12 cluster teachers, including Hattie Cotton Elementary School, Bailey Middle School, and Stratford High School, met with the instructional designers. They decided there had to be vertical continuity between the schools so that when the students arrived, they had a developing sense of big concepts through the themes of patterns, cause and effect, systems, and change, in that order. Kathryn Lee, instructional designer at Stratford High School, reports, "They were based on the AAAS 2061 themes and they are also in the National Social Studies themes, in the National Math Teacher's themes and the Next-Gen Science connecting concepts." Without some sort of system in which to record, share, and edit, the work accomplished will be lost.

The STEM shift in each school and district will be guided by individual plans. Implementation choices will vary widely. Resources and readiness levels will be specific to schools and to teachers. It can begin as extracurricular or cocurricular, in one classroom or across classrooms. But a systemwide shift is an organizational shift in the manner in which teaching and learning take place in all grade levels. In all cases, transsubject interdisciplinary integrated curriculum will become a fundamental practice and students will become more engaged learners. Parker Palmer (2011) observes, "The relational dynamics of the classroom have a more lasting impact on the students than information that they retain just long enough to pass the test" (p. 133). Scholars support the significance of the shift as real-world learning.

> STEM is particularly suited for PBL because of the natural overlap between the fields of science, technology, engineering and mathematics. In the real world, solving social and environmental problems does not occur in isolated domains, but rather at the boundaries of the STEM fields. (Capraro et al., 2013, p. 51)

REFERENCES

Capraro, R. M., Capraro, M. M., & Morgan, J. R. (Eds.). (2013). *STEM project-based learning: An integrated science, technology, engineering, and mathematics (STEM) approach* (2nd ed.). Boston, MA: Sense Publishers.

Bransford, J. D., Brown, A. L., & Cocking, R. R. (Eds.). (2000). *How people learn: Brain, mind, experience, and school.* Washington, DC: National Academy Press.

Palmer, P. J. (2011). *Healing the heart of democracy.* San Francisco, CA: Jossey-Bass.

Paul, R., & Elder, L. (2006). *The thinker's guide to the art of Socratic questioning.* Tomales, CA: Foundation for Critical Thinking Press.

RESOURCES

 Access live links at **http://bit.ly/TheSTEMShift.**

Buck Institute for Education (BIE). *Project Search* (for curated project-based learning examples): http://bie.org/project_search/results/search/P450

Eesha Khare Inventing the One Minute Mobile Phone Charger (video): https://www.youtube.com/watch?v=kMWWOnnS3ZM **(QR code on page 102)**

Mayo, G. (2014): http://learning.blogs.nytimes.com/2014/05/15/reader-idea-creating-architectural-models-of-literary-themes/

Metro Nashville Public Schools. (2015). *Stratford High School Profile*: http://www
.mnps.org/pages/mnps/About_Us/MNPS_Schools/High_Schools/High_
School_Profile/Stratford_High_School_Profile

Montgomery County Public Schools. *Vision, Mission, and Core Values*: http://
www.montgomeryschoolsmd.org/boe/about/mission.aspx

New Milford High School. (2014). *Program of Studies 2014–2015 Academic Season*:
http://www.newmilfordschools.org/NMHS/media/Course_of_Study_2014-
15.pdf

Pericoli, M. (2013). *Writers as Architects*: http://opinionator.blogs.nytimes
.com/2013/08/03/writers-as-architects/

President Obama Speaks at the 2014 White House Science Fair (video): http://youtu
.be/vke-SE1mqIs **(QR code on page 103)**

Project Lead the Way: https://www.pltw.org

Project Lead the Way. (2014). *The Louis Calder Foundation Puts Support Behind
Elementary STEM Education*: https://www.pltw.org/news/items/201405-
louis-calder-foundation-puts-support-behind-elementary-stem-education

Stratford High School Profile. http://www.mnps.org/pages/mnps/About_Us/
MNPS_Schools/High_Schools/High_School_Profile/Stratford_High_
School_Profile **(QR code on page 97)**

Stratford STEM Magnet High School (video with Principal Michael Steele): http://
youtu.be/4bWFb9vjEXE **(QR code on page 96)**

The Virginia Initiative for Science Teaching and Achievement (VISTA).*What Is
VISTA? A Program Overview* (video): http://youtu.be/5M3n3Vlfyog

10 Developing Capacity

STEM-Centric Professional Development

Human resources are like natural resources; they're often buried deep. You have to go looking for them; they're not just lying around on the surface. You have to create the circumstances where they show themselves.

—Sir Ken Robinson

High-quality, ongoing professional development is essential in order to inform and support the growth of the faculty as they incorporate new thinking, new practice, new information, new technologies, and reflection into teaching. We argue that all those entering a STEM shift also need an increased understanding of change and of trust building. In order to make this a systemwide sustainable shift in practice, teachers have to feel confident and safe enough to take risks that do not end up as negative performance judgments. All entering unknown territories need to be able to try and fail, learn, and try again.

In her study of the role of the superintendent in redefining schools for students to be successful in the 21st century, Sarah Stack Feinberg found that knowledge, skills, and abilities of teachers, leaders, and other staff members were dependent on professional development. Whether through formal, traditional professional development opportunities or the creation of Professional Learning Communities, she found that the **reinventing of schools for the 21st century demanded that professional development be** embedded into the culture (Feinberg, 2012). Changing the way learning occurs cannot be a mandate. Learning, or relearning, how students learn best and what the environment for that to take place requires, must be accompanied by ongoing learning, reflection, and discussion, with all engaged in the process.

Dr. Eli Eisenberg believes the quality of any educational system is no higher than the quality of its teachers. Without rigorous and focused ongoing learning opportunities, educators will teach the way they were taught. Those practices that worked in the traditional school setting will falter when a foundational shift is initiated. Those who were the best teachers may need to confront the prospect that others will find the new delivery system more comfortable than they and perhaps will outperform them. **Ego and identity crises lurk in corners as the shift begins.** Few human beings like having to begin again when they have mastered a career, but this is what a STEM shift requires. Why? Because we begin to search for problems and possibilities rather than facts and answers.

We are reminded of the wisdom of Shunryu Suzuki (1973), the Japanese monk who taught that **in the beginner's mind there are many possibilities,** but in the expert's there are few. Changing a practice that has been long a part of schools and has been socialized throughout the system is difficult. But if we want teachers, leaders, and parents to embrace the STEM potential, rich, valuable, ongoing professional development must be provided.

• BEGINNING THE DEVELOPMENT

Professional development begins in the moment when exploration of STEM schools and initial STEM conversations begin. It can spring from the most informal conversations among colleagues, school leaders, community leaders, parents, and students. The early musings reveal the thinking that exists in the organization. They provide the first opportunity to gather and share important information and unpack fact from myth. For example, in those early conversations, even those that are informal and off-the-cuff, one can hear acceptance and resistance, hope and fear, eagerness and hesitation. This data can allow leaders to ascertain the readiness of those who will be asked to join the shift process. That data will help shape the direction the formal professional development should take. The data gathering expands when visiting other sites, sharing observations and thoughts among visitation team members and with others upon return. Some schools and districts will collect readiness data from surveys in a more structured manner, but the early data all informs the leadership team and clarifies the professional development strategy.

One facet of professional development planning even can begin during the search for models. This became clear to Tyler Howe and his team.

Neil Armstrong Academy, Granite School District

While traveling around the country looking for models and collecting ideas for designing their Academy, Principal Tyler Howe and his team were searching for programs with an approach focused on the methodologies found in science, technology, math, and engineering. Their district already had a technical high school and now was planning to open a STEM elementary school followed by a STEM middle school, and then a STEM high school. Howe and his team wanted to see authentic opportunities aligned with Common Core Standards. They wanted to design a school that would have elementary students solve problems and grapple with ideas—a school in which students would discuss and present to their teammates and classmates and extrapolate from simple to complex concepts. Howe described the philosophical shift regarding the manner in which teachers would be encouraged and supported to take risks.

> Before opening Neil Armstrong, the staff of one of the schools we visited that had a STEM focus for a few years talked about the importance of their teachers and professional development. They believed that just like with their students, it is important to allow teachers to try something and have it not work and try it again. They encouraged taking that failure to heart and making tweaks along the way. I think that is so powerful, it became one of our collective commitments in our recent mission statement. We wanted to agree and value that if we're really expecting students to break from the mold of simply sit and listen to the teacher and we're expecting kids to be given the permission to think, try something out and maybe not have that work, go back to the drawing board, try it a different way, and then ultimately have success—then we need the teachers to feel that same liberty. And one of our teachers said to me, "Are you serious? With high-stakes testing and things like that, you're saying it's okay for us to fail along the way." And I said to her, "You know, I'm convinced it's okay because if we can really act like scientists do, we're going to have our best successes revealed on those big high-stakes public tests at the end of the year." We have to systematically go through our approaches—Did they get it? Did this work? If not, what else are we going do? How are we going to respond to kids who already got it? If we're not systematically doing that and feeling free to fail along the way,

then the only big failure we have is the very end of the year—the high-stakes tests. I think we needed to be patient on both ends with the kids and with the teachers. This change is a little bit flying in the face of years and years and years of past practice. Teachers and students have developed within a paradigm we have to change. It cannot be broken down overnight. (personal communication, February 28, 2014)

Shifting to any new practice has its best hope for success when training comes from outside experts and is reinforced locally. Reflection and sharing are components in every session. Opportunities for professional seminars allow thinking and talking about the shift that will take place or is taking place. This helps improve individual practice and allows for peer feedback among colleagues.

> Ideas and proposals most certainly can come from within our schools, but unless we pollenate our thinking with expertise from outside our walls, we may find ourselves spinning our wheels or using scarce resources redundantly.

Beware that changing practices in teaching will also involve hesitation; it may give rise to cynicism and to judgment. The brave few who speak out early in support of the shift will possess a great deal of courage. Understanding the change process will help everyone identify where colleagues are along the way and what support they might need.

Districts struggling with dwindling resources, reduced workforce and academic offerings, shrinking athletic opportunities, and perhaps even increasing class sizes may find the idea of investing in a STEM shift and professional development unthinkable. Professional development and coaching provide valuable support as teachers and school leaders are asked to consider, openly, the opportunities presented by interdisciplinary problem- and project-based learning.

Segments of professional development might take place in learning communities within the school or district. Other aspects will be required to be provided by experts outside the district or school. In making these decisions and creating professional development plans, teacher leaders should be invited to offer guidance. Their input should be trusted. It is one way environments of relational trust can be built. As professional development plans are created and resources allocated to fund them, remember that each phase and each new practice will need development of some sort. This is true from creating transdisciplinary lessons to working with STEM professionals from the field. Roles and relationships will be new. Accountability systems will be also.

● RESOURCES AND PARTNERS

There is another new aspect of the STEM shift that has a positive capacity impact. Community partners are eager to join the effort and support schools that want to make the shift by sharing resources, assisting with professional development, and participating as colleagues in problems and projects for students.

Schools entering a STEM shift will discover that prolific resources exist as well as many organizations with the express purpose of supporting this shift. Affiliation examples include with organizations like Public Impact in Chapel Hill, North Carolina, and Project LIFT in Charlotte, North Carolina (see Resources). The National Education Association (NEA) has developed a resource bank for local educators and provides STEM training programs and connects teachers to NASA programs, to the National STEM video game challenge, and to the SMILE website for lessons (see Resources). There are STEM hubs across the country as well as universities, arts councils, businesses, and hospitals. All have resources to provide and partnerships that can enhance the development of a program through intensive, targeted professional development.

East Syracuse Minoa Central School District

The East Syracuse Minoa (ESM) middle school initially began the shift in earnest, impacting the elementary school and the high school in tangential ways. A learning tour to schools that had already made the STEM shift ignited important planning discussions even before ESM personnel returned home. The visiting team was excited and convinced and began to talk about how they wanted to implement STEM back in their schools. During the spring and summer that followed the fall learning tour, they began to work with people like John Barell, who wrote *Why Are School Buses Always Yellow?: Teaching for Inquiry PreK–5* and *Developing More Curious Minds*.

Barell came to ESM and conducted a workshop on inquiry for a cross section of teachers in elementary, middle, and high school. They also had him present an evening workshop for parents and educational leaders across the district. Their intent was to develop a common, shared knowledge about inquiry-based learning. They wanted to be sure everyone understood what changes when students are solving problems that require curiosity, inquiry, and a higher level of rigor, rather than, "I need to know this in order to pass the test at the end of the week." The shift to what a student needs to know in order to solve a real-world problem required understanding across the community: parents, educators, and leaders in the district. They also worked with Dr. Robert and Rebecca DuFour on building professional learning communities. With the focus brought by these professionals, ESM was on its way to building capacity as a professional learning community and as an inquiry-based learning district.

DuPont Hadley Middle School, Metro Nashville Public Schools

When Tennessee applied for and won a Race to the Top (RTTT) Grant of $501 million, they were sure to include professional development opportunities through partnerships formed with the Tennessee Higher Education Commission and Battelle Memorial Institute.

The following is from the Tennessee Department of Education (2014) website:

The Tennessee Higher Education Commission and Battelle Memorial Institute

Link also available at http://bit.ly/TheSTEMShift

> These collaborations provide support for STEM (science, technology, engineering, and math) professional development for K–12 teachers, the College Access and Success Network, more effective teacher preparation programs, and enhanced stakeholder engagement and collaborations that bring together business and education to motivate teachers and students alike.

So in Tennessee, along with other RTTT initiatives, STEM education and project-based learning with a guarantee of professional development was incentivized. From this work came the establishment of STEM Innovation Hubs. DuPont Hadley is not one of Metro Nashville's magnet STEM schools but has embraced project- and problem-based learning with the support and encouragement of the Middle Tennessee STEM Innovation Hub.

Pamela Newman, one of the fifth-grade teachers on the team, said the following:

> We are not a STEM school, but if we're talking about integrated curriculum—science, technology, math, and engineering—that's the way our world is. We don't live in little compartments, and everything is not isolated. All of those subject areas intermingle and interact with each other. So it's just natural. It's a natural way to learn. (personal communication, 2014)

Buck Institute
for Education
Project
Search

Link also
available at
http://bit.ly/
TheSTEMShift

According to Newman, all teachers in the district attended project-based learning (PBL) training done by the Buck Institute of Education. The Institute's website defines *project-based learning* as "a teaching method in which students gain knowledge and skills by working for an extended period of time to investigate and respond to a complex question, problem, or challenge." This website is rich with free information, rubrics, and a **treasure trove of projects** that offer frames of reference for both beginners and experienced PBL teams.

The training for DuPont Hadley teachers took place over the summer of 2013, with an associated expectation for lesson implementation that fall. The school had embraced the belief that to prepare students for adult life in the 21st century, project-based learning was essential. Faculty, including the fifth-grade team, knew they needed to learn the model. The Buck Institute required that when the teachers left, their project be designed and ready to be implemented.

The teachers from DuPont Hadley built the framework, and the curricula and standards were integrated. They intended to implement a project-based integrated unit on the study of cells. A fifth-grade student had been diagnosed with cancer and shared her situation with teachers and students in the school. Teachers reported that the momentum in this first school year was supported by the relevance of the project to this student's life and to the student's classmates. The opportunity for the authentic learning this presented fueled their work during that summer. So with permission of the student and her parents, the study of cells became centered on this child's cancer diagnosis. In fact, the whole fifth grade, over 170 students, gathered in the fall to listen to their classmate describe her cancer experience.

> Students worked with scientists and doctors from Vanderbilt University Hospital, learning about pediatric cancer. The Center for Science Outreach at Vanderbilt brought live cells and different kinds of microscopes, equipment that we did not have in our schools. Without them, students would be not have been able to see the paramecium and look at different kinds of cells, live cells. It made it real for them. And, the early results in fifth-grade math went up from 20% to 52% proficient at the grade level. (Pam Newman, personal communication, 2014)

The students wanted to support Vanderbilt University Hospital and planned Cell-e-brate. It was both a fund-raiser and an opportunity to present the findings of their work on this unit of study to the public. Parents, scientists, and doctors from Vanderbilt University Hospital attended.

K–12 Instructional Designers:
Metro Nashville Public Schools

In Metro Nashville, instructional designers worked with teachers and students in designated STEM magnet schools, designing curriculum and modeling teaching in the classroom, basically coteaching and then serving as a coach as implementation grew stronger. The instructional designers were paid from a Magnet Schools Assistance Program (MSAP) grant from the U.S. Department of Education. While the designers served a particular school, they also worked as part of a K–12 design team to ensure vertical alignment as students moved through the system.

Kathryn Lee is one of the instructional designers at Stratford STEM Magnet High School. Following the implementation of lessons, the instructional designers work with the teachers to analyze the successes and the needs for changes that may be indicated. This feedback dialogue between the instructional designer and the teacher is conducted in the spirit of continuous improvement. The grant required an articulated plan for sustainability. When the grant ran out, the district was committed to supporting the instructional design role and process moving forward.

For the upcoming year, Lee will be working half time with the teachers of AP Chemistry and AP Biology classes to repeat the process of designing curriculum, modeling with them in the classroom, coaching, reflecting, and adjusting the plan. The goal is for the instructional design team to bring the faculty to a point from which STEM transdisciplinary and project-based learning is solidly established and at which the instructional designer is no longer needed.

LaKisha Brinson, presently Library Media Specialist at Robert E. Lillard Elementary School, was formerly an instructional designer at Hattie Cotton STEM Magnet Elementary School. She observed,

> It is my belief that becoming a STEM school caused a remarkable shift in the culture among the staff. A risk-free environment was established by the end of the three-year period, and teachers began to take ownership of their own project planning and experiences. The multitude of professional development opportunities offered increased teachers' content knowledge and provided numerous opportunities for teachers to collaborate at higher levels than before. (personal communication, May 29, 2014)

Ranson IB Middle School, Charlotte-Mecklenburg Public Schools

Ranson International Baccalaureate (IB) Middle School in Charlotte, North Carolina, offers the IB curriculum schoolwide. Though not a STEM school, we discovered an example of professional development (PD) through collaboration and innovation that merits inclusion. More than 75 percent of the Ranson Middle School students receive free or reduced lunch, and it is a Title 1 School. According to multiclassroom leader Romain Bertrand, part of the reason for the school's poor results in math and science scores was the constant turnover of teachers. They relied heavily on Teach for America teachers who would come and stay for a short while and then leave. In 2011, the school had negative growth in math scores, certainly not meeting growth expectations.

The school partnered with Public Impact and Project L.I.F.T. (Project Leadership & Investment For Transformation), a public/private partnership organized as a nonprofit organization. Their work directly impacts nine schools in the Charlotte-Mecklenburg School System. Project L.I.F.T's mission is to "transform the way students who traditionally perform poorly in school are educated by ensuring these students are equipped and ready to enter the 21st century and beyond." Their intent is to strengthen students' educational foundation using a unique approach to teaching and learning that includes bold, innovative strategies.

Mr. Bertrand was engaged in a leadership program called SELA (School Executives Leadership Academy), a residency-based leadership program at the Queens University of Charlotte. After completing five weeks of intensive classes in the summer internship program, he was to do a residency at Ranson. He had to choose a topic of research in a residency project that would have an impact on student achievement. At that time, the school was beginning to use what is called *rotation models*, using blended learning that allows teachers to better differentiate instruction and better serve the needs of their students. Their class size for math was around twenty to twenty-five students. Once lab rotation and classroom rotation models in sixth and seventh grades were started, class sizes in math grew to close to thirty. However, on any given day during a math block, a math teacher would work with a small group of fifteen, in face-to-face instruction, while another fifteen worked online. Then they rotated. In order to meet the highly differentiated needs of the students, Mr. Bertrand reported, "We found some of our students were helped to catch up while others were totally ready right away for more rigorous instruction at a much faster pace. Our goal was that both groups would be met with the same level of rigor."

Mr. Bertrand describes their present approach to be more traditional than he would like to see it moving forward. Presently they are utilizing block times through which they create the rotations. He questions whether this model is the end of the shift for them or one to use on the journey to an even more improved end. But what keeps him and his colleagues on this track is that they think, for now, it is helping them build something that will allow students to come into class and never be bored, never feel slowed down, never feel trapped in a lesson that they already know or that is too hard for them. Mr. Bertrand recommends, as they did for their blended learning model, "I encourage people to have a clear vision, an inspiring vision. Build it gradually with steps that allow people to learn the basics and test it out. Celebrate the victories and feel more confident so that then they want to do it more."

Charlotte-Mecklenburg Schools Strategic Plan 2018

Link also available at http://bit.ly/TheSTEMShift

The sixth of their six strategic goals reads "Inspire and nurture learning, creativity, innovation and entrepreneurship through technology and strategic school redesign." The district may be inching toward a STEM shift. It included all of its schools in a week of coding. There are a growing number of clubs and projects, including technology and robotics. And, according to Mr. Bertrand, the district is trying to figure out how all of this could be better included into the curriculum.

● COACHING IS ESSENTIAL

Teachers must be learning, supported, encouraged, coached, and accountable. Mentoring and coaching have similarities, but they are clearly different. Mentoring is usually provided to those at the entry level, either to a position or to a profession. Individuals at all stages of their careers benefit from coaching. Athletes and musicians know that; educators, not so much. But it takes work to stay at the top of your field and on the cutting edge. Coaching is a facilitative, not a directive, process. The success of any coaching program depends on trust and confidentiality. Coaches need to be carefully chosen and carefully trained.

There are many coaching models and processes circulating now. One focused purposefully on educators is evocative coaching, designed by Bob and Megan Tschannen-Moran (2010). Authors of the book *Evocative Coaching: Transforming Schools One Conversation at a Time*, they say, "Coaching is a process that brings out the greatness in people. It raises the bar of the possible so that people reinvent themselves and their organizations" (p. 5). This is precisely what STEM-shift schools seek.

What Is
VISTA?

Link also
available at
http://bit.ly/
TheSTEMShift

George Mason University

One example comes from the Commonwealth of Virginia at George Mason University's Center for Restructuring Education in Science and Technology (CREST). They run the Virginia Initiative for Science Teaching and Achievement (VISTA), an i3 grant from the U.S. Department of Education.

VISTA provides scholarships for professional development opportunities to science teachers and leaders with the goal of transforming the teaching and learning of science in Grades 4–16+.

The following is from VISTA's website:

> The Virginia Initiative for Science Teaching and Achievement (VISTA) is a statewide partnership among 60+ Virginia school districts, six Virginia universities, and the Virginia Department of Education. Its goal is to translate research-based best teaching practices into improved science teaching and student learning for all students at all levels.

VISTA
Website

Link also
available at
http://bit.ly/
TheSTEMShift

Their goal is to improve science teaching and achievement. An overview can be seen on YouTube.

In addition to the learning, practicing, collaborating, trial and error, reflection, and redesigning in this change process, the role of a coach in important. It is through VISTA that we met fourth-grade teacher Beth Ferguson, and through her, the Goochland County Public Schools. VISTA provided an intensive five-week summer course for science teachers. Following that learning experience, those teachers were assigned coaches. Beth Ferguson shared that

VISTA
Voices

Link also
available at
http://bit.ly/
TheSTEMShift

> It's great to have a coach. She comes and meets with us on a regular basis. We have dialogue with her all of the time. She's made us look at how we are teaching science specifically, which is what her role is, really. She makes us think about the nature of science and all of its tenets, and taking it from the science perspective and really moving it across the curriculum. We already felt pretty comfortable in planning and teaching cross-curricular subjects, but it gave us that avenue to really think about how we can be better science teachers by asking better questions and pulling more from the students. We've taken all of that that and applied to the other subject areas and incorporated it all together. (personal communication, January 30, 2014)

Affiliations with entities beyond school building walls, like the one with George Mason, that are prepared to offer long-term, focused professional development and coaching are the best bet for initiating a STEM shift in these difficult financial times.

● EMERGENCE AND CONVERGENCE

Excellent teachers know how to teach and help children learn. If we allow them the space to take risks, letting go of traditional instruction grasping this integrated notion of experimentation and exploration, the shift will be successful at the teaching learning point of contact. In Romain Bertrand's classes, a route to improved mathematical thinking and skills is being built through the use of blended learning for the students, with ongoing, embedded professional development for the teachers. As those skills grow and students catch up and move ahead, and as the science department investigates blended learning to achieve the same thing, the interdisciplinary nature of those subjects will emerge organically. In Tyler Howe's school and those schools in Metro Nashville and East Syracuse Minoa, the integration of curriculum has already shifted to problem- and project-based learning systems. These educational pioneers are engaging children in learning, even those children who previously seemed unengaged.

Dr. Eisenberg describes the STEM shift as moving from only being concerned with knowledge to creating know-how. He believes many teachers are prepared to take these risks and make the shift while others will accept the shift more slowly. All will require support and training. **The shift can be successful only if the teachers are prepared and supported to move from the familiar into the unfamiliar, from silos to collaborative work.** Knowing how means knowledge and skill and willingness. Implementation becomes the stage on which to demonstrate understanding and explore possibilities. We all learn better when the application is real, rather in the abstract. It is true for teachers and for students.

No matter the country, the understanding of a STEM shift is commonly shared. It is a K–12 shift that involves all areas of study, not just the STEM areas. It is a change in the way things are taught, a change in the way we do business, a change in the way we provide professional development, and a change in resources and an opening up for new partners. In example after example, we found partners ready and willing to join in the shift when invited. It will take some searching and relationship building. It may take grant writing. But excitement is being generated because of the school-business collaborations and the teacher-STEM professional partnerships. Schools are becoming part of the public again, a hub of activity for all, not just parents and students and teachers. The STEM shift is inclusive and extensive, local and worldwide.

In order to successfully make a STEM shift, affiliations will influence and support professional development, and they can and should help us teach. Business, the health industry, colleges, and universities must become our partners as we teach in an authentic transdisciplinary STEM-shifted environment.

A STEM System

At East Syracuse Minoa, the Spartan branch of the local credit union (CORE Federal Credit Union) is located in the lobby of the high school. It is led by the students who manage the accounts. Students have to develop résumés and interview for jobs. Once hired, they have to do the marketing for that branch. They are responsible for the hiring and the running of the branch. This is all done with the cooperation and support of the CORE Federal Credit Union and the district's teachers.

Superintendent Donna DeSiato reflects, "This really does create a shift in the way we are looking at what teaching and learning needs to look like in the 21st century." Students have the opportunity to do research with Bristol-Myers Squibb in which they engage in researching a new pharmaceutical drug. Dr. DeSiato explains:

> At the high school, where it is a little bit more challenging to do the transdisciplinary models, we actually have several different models which we have developed. One was brought to us by Bristol-Myers Squibb when they heard about what we were doing with our strategic plan. It is called Research: An Educational Journey. It is the study of the development of a pharmaceutical drug. It is co-taught by a teacher certified in math, one certified in biology, one certified in chemistry, a teacher who is dually certified in math and science and a teacher certified in business and marketing. That curriculum was developed collaboratively with the Smithsonian Institute and National Science Resource Center and given to our teachers for the last five years. We teach this credit-bearing course an hour before school begins over the course of many months. They come into it with the problem that the high school has developed a disease for which there is no known cure. You are a team of scientists who are now going to be researching what type of antidote or what do we need to develop to address this particular disease. It has some humor to it. It is called High School Disease. It has a set of symptoms. Students learn a tremendous amount about the FDA approval

process, the drug development process, and we see it as a shift because we're no longer teaching the science, the math, the business, the marketing, the ethics all in isolation. They are taught in the integration of the course because the course . . . Students are never late and never absent because they are so engaged in that learning. (personal communication, June 6, 2014)

Each year, that class tours the facility at Bristol-Myers Squibb at the culmination of their coursework. As with the opportunities given students at Stratford STEM Magnet High School in Metro Nashville, students present to the research scientists and to their parents. The students from East Syracuse Minoa are invited for the tour of this highly secure facility in the afternoon with their teachers, and the parents are invited with the research scientists in the evening for dinner and for the presentation.

In the middle school, the STREAM TEAM did a PBL unit called Back From the Dead: Onondaga Lake Plan. Onondaga Lake is highly polluted. Honeywell was the team's partner in this research. Students did an actual field visit to lake and collected all kinds of data; they had to develop their project around four different aspects. Then students presented their findings as the culmination for their particular unit.

In the eighth-grade We Built This City unit, engineers from Siemens and the architects from King and King come in and work with the students. As Dr. DeSiato explains,

The 8th grade students were all fired from 8th grade. They all had to apply and they are being hired as either engineers or project managers or the people on the teams that are studying nine different alternate sources of energy. There are members of the 8th grade that will be in the role of the village council. So there will be the mayor and the trustees, and in June, the students in each of those 9 energy research teams will be presenting their model and their findings for why, whether it is solar energy or hydro-electric energy or wind energy . . . whatever it is that their team has been studying why they would propose to the village that this be the source of energy that the village would use going into the future. We want in order for students to truly be prepared for the jobs that we know and the jobs we have yet to understand that will be there in 5 years, 10 years, 20 years from now. (personal communication, June 6, 2014)

Historically, educators have worked to include, improve, and increase the capacity of our classrooms. Our walls are bulging with programs that have been added in response to the needs of our students as they have increased and diversified. This shift requires us to rethink and repurpose what is within those walls and make them penetrable. Partnerships for STEM initiatives lie outside those walls in small businesses, nanotechnology manufacturing complexes, the health-care industry, and higher education. These affiliations help provide professional development for our teachers and leaders. They extend professionals to work alongside our faculty and facilities and equipment for collaborative work so that our students experience the authentic opportunities a STEM shift requires. They build the capacity of the educational system and help us communicate that 21st century teaching and learning have arrived.

REFERENCES

Feinberg, S. S. (2012). *The role of the superintendent in redefining schools for students to be successful in the 21st century* (Doctoral dissertation). Retrieved from http://library3.sage.edu/archive/thesis/ED/2012feinberg_s.PDF

Suzuki, S. (1973). *Zen mind, beginner's mind: Informal talks on Zen meditation and practice.* New York, NY: Weatherhill.

Tschannen-Moran, B., & Tschannen-Moran, M. (2010). *Evocative coaching: Transforming schools one conversation at a time.* San Francisco, CA: Jossey-Bass.

RESOURCES

 Access live links at **http://bit.ly/TheSTEMShift.**

Buck Institute for Education (BIE). *Project Search* (for curated project-based learning examples: http://bie.org/project_search/results/search/P450 **(QR code on page 112)**

Charlotte-Mecklenburg Schools. *Strategic Plan 2018*: http://www.cms.k12.nc.us/mediaroom/strategicplan2018/Documents/Strategic%20Plan%202018%20For%20a%20Better%20Tomorrow%20Fact%20Sheet.pdf **(QR code on page 115)**

DuPont Hadley Middle School: http://duponthadleyms.mnps.org/pages/DuPont_Hadley_Middle_School

Middle Tennessee STEM Hub: http://midtnstem.com

National Education Association. *The 10 Best STEM Resources Science, Technology, Engineering & Mathematics Resources for PreK–12*: http://www.nea.org/tools/lessons/stem-resources.html

Project LIFT. *About Project L.I.F.T.* (public/private partnership nonprofit organization, operating as one of five learning communities in the Charlotte-Mecklenburg School System): http://www.projectliftcharlotte.org/about

Public Impact (Helping education leaders and policymakersimprove student learning in K–12 education). *About Public Impact*: http://publicimpact.com/about-public-impact/

Tennessee Department of Education. (2014). *First to the Top*: http://www.tn.gov/education/about/fttt.shtml

The Tennessee Higher Education Commission and Battelle Memorial Institute. *Battelle Memorial Institute* (video): http://youtu.be/8PckHDH_6Ho **(QR code on page 111)**

The Virginia Initiative for Science Teaching and Achievement (VISTA): http://vista.gmu.edu **(QR code on page 116)**

The Virginia Initiative for Science Teaching and Achievement (VISTA). *VISTA Voices* (videos): https://www.youtube.com/user/VISTAScience?feature=watch **(QR code on page 116)**

The Virginia Initiative for Science Teaching and Achievement (VISTA). *What Is VISTA? A Program Overview*: http://youtu.be/5M3n3Vlfyog **(QR code on page 116)**

11 Time for STEM

> *A pair of scientists won the Nobel Prize in physics for isolating a material called graphene, the thinnest, strongest, most conductive material in existence. They did this work during what they called "Friday evening experiments" . . . making one of the greatest breakthroughs in the last 50 years, basically during a physicists' recess.*
>
> —Daniel Pink

Lessons will not be contained in thirty- to forty-minute chunks. How could we have professionally held for so long to the belief that young people could learn best with time limitations like that? Going forward in a STEM-shifted school, lessons will be based on problems and will raise questions. Time will serve rather than dictate and divide. Time will allow for independent learning and work in teams. The learning experience will include how to work together as a team and how to analyze problems, methods for research, discovery of resources and results, and presentation of those results. In many cases at all grade levels, STEM professionals from the community and around the world via Internet will be involved with the student learners. Yes, even in primary classrooms! This is the emerging work of the children.

● SCHEDULES ARE CURIOUS BEASTS

Few educators are well trained or expertly skilled at building and rebuilding schedules. Even fewer enjoy those tasks. It is typically a burdensome activity assigned to a less than excited administrator. The "rollover" schedule was most appealing. If it was working for most, "Don't disturb it" was the mantra. In many high schools across the nation, those who were preparing to enter a career or a two-year career preparation program upon high school graduation left the school for half of the day to attend offsite vocational programs.

> Time will serve rather than dictate and divide.

Schedules for the core courses had to accommodate those students. Access to gyms and labs need to be considered. Teacher recommendations for student placement also require consideration. The old model appeared to work for adults and for students, so it has been repeated for decades. Elementary-level scheduling is not the behemoth it is on the secondary level. However, deconstructing the elementary schedule and starting from scratch may also be a daunting task. It can also be extremely helpful in unexpected ways.

At the outset of a STEM-system shift, transparency demands that everyone know what predictable past practices may be impacted by in the process. Time is one. Anyone who has approached schedule changes in the past knows that these are highly contentious discussions. Yet, a discussion centered on priorities, brain research, developmental phases of children, and learning is required.

All voices need to be heard, but those leading the conversation might better be the experts and the most passionate supporters of STEM's potential than the administrator with the highest rank. Time constructs must be opened, not for a whim or a contract leash snap, but for the shift to maximize its potential. The discussion needs to stay in the bigger context. It isn't about changing the schedule. It is about using time as a resource in service of a new learning model.

Open, honest, creative dialogue, in which both concerns and opportunities can be raised, will help launch the shift. Input from all teachers in the system can become part of the conversation by the use of surveys and invitations to the planning conversations. This opportunity should be extended even to those whose own setting is not in the first

> The STEM shift requires using time as a resource in service of a new learning model.

phase of implementation. Visitations, virtual or real, to schools and systems a few steps further into the shift can be valuable. STEM professionals and visits to STEM worksites will demonstrate the need to open up time. The obstacles created by the present schedule need to be identified. **No fundamental shift fits neatly into an old container.** STEM requires implementing and experimenting with subject integration, and that will change teaching practices.

In large systems, finding willing participants and determining building sites may be easier than in small systems where options are fewer.

Nevertheless, the process itself involves the same change leadership practices. Courageous conversations happen. Everyone becomes a novice in these early stages. When boldly stepping into the schedule issue, it is natural to brace for loss responses. The ones leading these conversations must be prepared to be empathic and compassionate leaders who can hear the uncertainty of those being asked to change their practice and yet not be deterred. Logistics of time cannot supersede what is the next right answer for students. Remember there will be some on every faculty eager to be engaged in this thematic, interdisciplinary, problem- and project-based learning practice. They need to be prepared to lead in a way that invites their colleagues to join them or follow them willingly.

Here, change expert William Bridges (2009) offers some thoughts:

- Allow for the identification and expression of behaviors and attitudes that will have to change to make things work.

- Identify who may lose something under the new system.

- Remind everyone of the reason for the change.

- Reach out to those troubled by the change.

- Continue to talk about transitions and how people go through them. (pp. 15–16)

● SPACE FOR ADULT LEARNING

A major shift in teaching and learning requires teacher training and time for collaboration. Without prioritizing time for a cycle of learning, trials, reflection, collaboration, redesign, and trying again, it is not likely to happen. As a resource, time to prepare is as essential as time to deliver. Create the time and the mental spaces for a shift in the way students are taught. Create the time and space for the faculty to share what they are learning as they are provided the opportunities to learn from experts outside of the building and district and from each other. Create the time and space for those courageous conversations about doubt, concern, hesitation, and success to take place. Learning is a constant in a STEM-shifted school. Students and their teachers will find themselves in a five-step learning process. Part of the shift involves allowing this cycle greater transparency as it becomes not just an individual experience but a group one as well (see Figure 11.1).

There is a relationship between time and space. As you enter a system shift, it is important to remember that shifting artificial time allocations is really related to the space within the system. All three dimensions of space—up/down, left/ right, forward/backward—will be impacted as a

Figure 11.1 Collaborative Learning Time

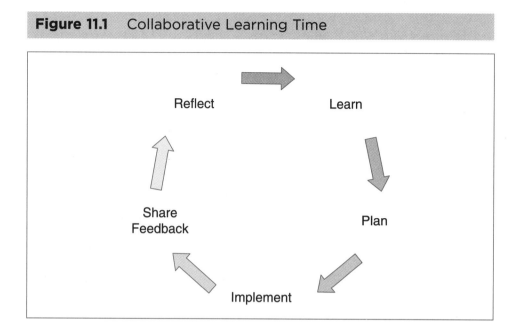

system shifts to STEM. If we begin even within the current limitations of school day and year, the space of six, seven, or eight hours a day is available for repurposing. The focused, purposeful dialogue about that time and space and how they are used can summon innovators to think newly.

The East Syracuse Minoa Central School District embraced the 21st Century Skills learning model in its strategic plan and has transformed the middle school. Dr. DeSiato observed that

> Initially, the schedule was redesigned to provide one of the 8th-grade teams with an uninterrupted block of 120 minutes, with students no longer following a bell-driven schedule. Learning is organized with the focus on the purpose of the lesson rather than on the predetermined number of minutes in a given period. Student learning groups are designed by engaging students in problem-based learning and determining appropriate group size. The team of teachers collaboratively organize learning groups that vary in size; they engage all 110-plus students on the team in a large-group interactive, collaborative lesson such as "morning meeting," or they divide students among the teachers on the team in groups of 25, or at times they vary the group size with 30 to 40 students or 8 to 10 students, depending on the focus of the learning. During the first year, as the team implemented the learning model with flexible, adaptable schedules and varied group size, there was a shift to learning becoming the constant and time becoming a variable. As

we hosted learning tours of teachers and administrators from other districts, there were many questions about how to organize and schedule a transdisciplinary learning model. Throughout the year, we continued to build capacity and deepen our understanding with ongoing professional development, research, application of our own professional learning, and by reflecting on our practice. (personal communication, June 6, 2014)

● FLEXIBILITY

The ease of embracing change for oneself varies among individuals, but it is always a process. Leading a system through a foundational shift magnifies that process. The expectation for a schedule to remain static is unrealistic. As opportunities expand for students to learn in multifaceted settings, as Dr. DeSiato describes, and eventually even outside of the school, the schedule will adapt. Once a system breaks from historical constructs, mental models shift. Then, operating under the strength of a core set of values, the process becomes one of discovery and flexibility. As everyone learns, reflects, and develops as member of a STEM shift, adjustments and corrections in course are expected. The schedule reflects the design and will be modified as a result of analysis, reflection, and input.

Elementary Schedules

Elementary schools differ in how they schedule their day. Schedules will vary greatly, depending upon the size of the school or district; the size of the staff; the requirements for the manner in which remediation, ELL services, and special education services are offered; or how instrumental and choral lessons are held.

Goochland County Public Schools in Virginia are several years into a STEM shift. The philosophy in their elementary school, for example, is to integrate hands-on, inquiry-focused design and delivery all day long. So the schedule is committed to allowing as much of that to happen as is possible. A commitment to such a philosophy sets the target and opens the door.

At Goochland, teachers work with the principal and those who provide support services to be sure all students receive the services they require, while not missing the learning opportunities with the rest of their classmates. This is of particular importance because much of their problem-based, project-based interdisciplinary work depends on all members of their team. Communication, juggling, and prioritizing as a team, they work to preserve the time for support services students require. This also includes music lessons and remediation. In cases with students who need to double or triple up on services, issues are discussed and the team works diligently to be sure the students' needs are met and their learning time is not interrupted.

It is not unusual in traditional scheduling that one person, or a small group, is in charge of these decisions. The contract, teachers' schedules, perceived priorities, and personal preferences all are part of a tug-of-war until the details of the schedule are finalized. In Goochland County Schools, Beth Ferguson, a fourth-grade teacher at Randolph Elementary School, and her team work together developing the schedule. The district's and school's priorities are clear and are the standard for decisions being made. Systemic approaches to interventions, extracurricular engagement for all, personalized learning plans, and STEM and CTE expansion are ingredients considered as they make their schedule. Those ingredients are among the priorities served by the leadership and the faculty's commitment to making the core values the foundation of every decision, large or small.

Secondary Schedules

The challenges for beginning the change in secondary schools are different. Secondary teachers, by training, tend not to be natural integrators. Most became secondary teachers because of a love of subject matter and an eagerness to explore that field with young people old enough to engage it. Their education is based mostly upon content knowledge. They are held accountable for teaching and assessing students' acquisition and application of the subject area. Generally speaking, supervisors observe and evaluate secondary teachers on their teaching and assessment of student learning of a subject. Even if there are literacy components included in the sciences or social studies, the focus on how to develop reading or writing capacities in those subjects is given little attention compared to the attention given to the topic or subject. This is true even though the Common Core requires a more intentional teaching of literacy across the curriculum.

The teachers of lower grades always have the hope that what they were not able to accomplish may be accomplished the following year, with new teachers in a new grade. This doesn't exist for the high school teacher. There is a sense of urgency among high school educators. If a student cannot master the material required in a course, there are not many years, or much space left, in which to repeat and then catch up in time for graduation. In high school, students use cumulatively the skills and information that they have acquired in elementary and middle school. High school can be an exciting place. It is also the end of a chapter in the book of their life.

It is not uncommon for subject departments in secondary schools to join together as advocacy groups. Departments and grade-level groups often vie for prominence in the schedule and value in the academic hierarchy of the school. Placement of courses within a schedule determine what is available and to which student subgroups. In worst-case scenarios, course placements determine teacher job security.

Within secondary schools, then, the starting place is to move away from silos in which teachers have become successful specialists in their subject areas. A step into transsubject organization, combining classes like trigonometry and physics; history, art, literature, and architecture; and math and technology begins the turning of minds and of practice, but can also set resistance into motion. When there are uniform tests at the end of courses, like the Regents Exams in New York State, the idea of creating a transsubject integrated course with perhaps two or more teachers or an expert contributing from the field, using problems and projects to help students synthesize their learning and apply their knowledge, and ending with an external assessment, is confounding.

Once interdisciplinary courses are created on the high school level, a successful program must include relationships with business, health care, colleges, and universities. These are the places where students have opportunities to work with professionals in the field. Professional experiences require students to spend learning time outside the school building. This calls for another major innovative step into creative scheduling and reconsideration of school day parameters. We must also question whether we have allowed adult needs and interests to supersede student ones as schedules are formed. STEM-shifted learning environments require a reversal of any of those practices. An example from Metro Nashville can illustrate the value of more open thinking.

Stratford STEM Magnet High School, Metro Nashville Public Schools

At Stratford STEM Magnet High School, students were allowed to leave for a full day once a week to attend classes at Vanderbilt University. Students in Grades 9–11 had one day each week (a different day for each grade level) when they were scheduled to be at Vanderbilt all day. On that day, they would take classes and work with scientists in the university's labs. Many students put extra time in the labs, working with scientists on their own time, to complete the projects that were undertaken. By the time they were seniors, many of the participating students had accumulated enough credits in their high school program to allow them to spend two full days a week working on individual research projects in laboratories at Vanderbilt.

When this program was brought into the Stratford STEM Magnet High School program, the scientists came to the school instead of having the students leave to go to Vanderbilt. The scientists now work alongside the high school science teachers at Stratford to coteach the Interdisciplinary Science Research (ISR) class at Stratford. Since the school elected to go to a modified block schedule, students are able to "double block" classes. They can have three hours every other day to work in the ISR class. The scientist-teachers are then able to teach four classes, one for each grade level, on that same schedule (Double block class 1 and 2, class 3 and 4 on "A Day"; and double block Class 5 and 6 as well as class 7 and 8 on "B Day"). Because of their literacy focus, they instituted twenty minutes of literacy in every content area in the school. So, within each and every subject, twenty minutes of reading was included in the block, allowing for the ongoing development of literacy in each content area.

● SCHEDULE THE DESIGN

How the schedule works depends on whether the school is organized as a STEM-shifted school or contains a STEM academy or is focused on project- or problem-based interdisciplinary STEM-based courses. It is important to determine how the school will be organized, what type of program is developing, how much freedom from structure will the program be given in the day. Models from which to learn and adapt exist.

   As with all change, it is best to begin where there is the most possibility for interest, investment, courage, and innovation. If it does not support the vision, a minor modification in the current practice may occur but not a true shift in the way schooling has been done. In elementary school, if grade-level teams are going to work on interdisciplinary integrated learning, arranging for common planning times for those grade-level teachers is a priority. If step one is simply to integrate science and art, arrangements for those teachers to plan together must be accommodated in the schedule. Especially in high school, the opportunity to learn in the field with scientists, physicians, engineers, architects, and graphic designers will require serious change in the use of time. The bottom line is that the current schedule limits a STEM shift by establishing artificial barriers rather than creating possibility. Remember, the schedule emerges from the vision.

REFERENCES

Azzam, A. (2014, September). Motivated to learn: A conversation with Daniel Pink. *Educational Leadership, 72*(1), 16.

Bridges, W. (2009). *Managing transitions: Making the most of change* (3rd ed.). Philadelphia, PA: Da Capo Press.

⑫ STEM Collaborations and Trust

In systems of trust people are free to create the relationships they need. Trust enables the system to open. The system expands to include those it had excluded. More conversations—more diverse and diverging views—become important. People decide to work with those from whom they had been separate.

—Margaret Wheatley & Myron Kellner-Rogers

A 21st century STEM learning environment relies heavily on collaborative work by teachers and students alike. The degree of enthusiasm with which adults enter a shift with the magnitude of punctuated equilibrium will determine the energy of their followership and the ultimate experiences of students. If, as we think, STEM shifts can secure the future of public education by keeping it at the center of society and economy, then they are important in whatever degree a locality chooses. This is not about the notion of a leader whose vision will atrophy when he or she leaves. It about the external pressures that cannot be ignored intersecting with the desire for American children to be among the most highly achieving in the world, among the most innovative, and the most free. This book would be missing a key element in enabling a successful shift if this chapter were missing.

It is naive to assert that people know how to be trustworthy and want to be collaborative. Many human beings, for reasons of personal life stories and personality, are reticent to trust and would rather work independently. To encourage followership for a change initiative, someone in the system must be paying attention to the people in the process and focused on the relationships that will ignite a creative and generative spirit.

● TRUST AS A SHIFT ESSENTIAL

Amid the national agenda about Common Core, intense training is being provided in standards, assessments, and lesson development. Professional development, it seems, must connect to student test-taking skills and results. It is as if we have forgotten that all adults who have chosen to work with children are first human beings, themselves. Those humanizing qualities are not checked at the schoolhouse door—or they ought not to be. The teachers we remember were those who brought more than the facts in their heads to work. They brought the inner qualities of themselves to care, to believe, to inspire, to reach out with compassion and into the world of every child with respectful encouragement. They also bring vulnerability and their own life stories.

School systems have been reluctant to spend public funds to prepare faculty and leaders for a work environment that requires trust and collaboration. There seems to be some underlying belief operating that all adults come hardwired with these propensities and skills and they emerge when called upon. Or, more deeply troubling, there is a mind-set that diminishes these skills as only tangential to the work of obtaining better student results. Stephen M. R. Covey (2006), the personal and organizational development guru, referred to trust as the lifeblood of an organization. His book *The Speed of Trust: The One Thing That Changes Everything* makes the business argument that leaders can increase the speed of change and decrease associated costs if trust levels are high. Franklin Covey (global consulting and training leader) contends that *trust*, as a verb, is a skill that can be developed, and they have designed training programs to do just that. If the approach is too much sales or too corporate for some of our readers, there is hard research to support the importance of trust in schools.

Amid change of any sort, we ought not overlook or forget the research of Megan Tschannen-Moran. It is one of several pieces that provide a convincing argument that that trust matters. In fact, *Trust Matters* is the title of her book, recently republished in its second edition. Trust is defined as "one's willingness to be vulnerable to another based on the confidence that the other is benevolent, honest, open, reliable, and competent" (2004, p. 17). These five factors, so simply fundamental and so difficult to find sometimes, are critical to building relational trust. The

2006 study by Tschannen-Moran and her colleagues "demonstrates that a bridging strategy [from schools to their communities] provides a . . . powerful construct as schools seek to engage their parents and community members and increase student achievement" (Tschannen-Moran, Parish, & DiPaola, 2006, p. 410). To undertake a STEM shift, parents and community leaders as well as teachers and students must be walking into the new territory with us, discovering what is there, revitalizing what we bring, and beginning anew.

Anthony S. Bryk and Barbara Schneider (2002) were, simultaneously, conducting studies of relational trust and school improvement in Chicago, Illinois. They concluded that "relational trust is a core resource for school improvement." Integral components of relational trust are respect, personal regard for others, competence in core role responsibilities, and integrity. Bryk and Schneider further suggest that any substantial lack of these components will work against the building of relationships and thus hinder student achievement. The lack of trust can disable a shift, but a shift investigation creates a profound opportunity to build trust, even in previously fractured systems.

> These studies and our own experiences cause us to assert that the STEM shift will occur faster and with greater ease in settings where trust is high and where it is nurtured.

These studies and our own experiences cause us to assert that the STEM shift will occur faster and with greater ease in settings where trust is high and where it is nurtured. A community of trust and collaboration must be intentionally created, nurtured, and sustained to carry people through a fundamental shift. For some, learning how to lead in such an environment may need to be part of the process. Leading schools into an emerging organization is substantively different from leading in the status quo while tinkering on its edges. Even to consider one fundamental difference would be highly disturbing to some. Teachers must be encouraged to innovate, and failure must be embraced as a lesson along the way. Risk taking must be reinforced for trust roots to develop.

Especially in the current culture of accountability, this shift can be successful only if leaders' and teachers' risk taking is met with approval, support, and encouragement. **Risk taking is part of the professional learning cycle; it produces failures and successes.** It reinforces creativity and career growth. It doesn't end them. Trust is essential in all schools and relationships, personal and professional, hoping to grow and be productive. This is especially true for schools and districts that enter a STEM shift.

● CREATING TRUSTWORTHY WORKPLACES

The STEM shift requires teachers become new kinds of experts in learning. Problem- and project-based learning and teaching in an integrated curriculum design require new thought and new teaching skills. Their original training as experts in teaching curricula and in behavior management becomes the platform from which they must emerge purposefully and faithfully as facilitators of learning and members of a team. **Teachers and their leaders become learners again.** The STEM-system leader must encourage teachers in all grades across the system to take risks, while learning alongside those teachers, so feedback is specific, informed, supportive, direct, and honest. The receiver feels respected, and feedback can be helpful and motivate for development.

Goochland County Public Schools

All of this takes courage, a quality not often found among qualifications listed on vacancy announcements for leaders or teachers. How can this be accomplished in today's environment defined by the Common Core and the standardized tests required to measure achievement?

Superintendent James Lane of the Goochland County Public Schools took a courageous stand by making clear that he didn't want student achievement to be the only measure of the STEM-shift success. Dr. Lane, along with his board, his central administration, and his building leaders, agreed that the district was going to focus on student engagement. With the leadership focusing on the implementation of these new teaching strategies, a theme-based project-based STEM shift, they decided student engagement would be a locally selected key measure for accountability purposes. They found and administered a survey that gauged student engagement. "We care more that our students are engaged every day in learning and define it as more than time on task."

Dr. Lane, his board, his leadership team, and teachers agree that if students are engaged then learning will happen. In addition, they are implementing a process to measure student growth. They value each student's growth and want to know how far students are growing in a year as opposed to what level of achievement they have attained. To get to that place takes work. This work is done despite the required multiple state tests given in Grades 3–8 and in high school. In addition to those required assessments, 40 percent of a teacher's evaluation is based on student academic progress.

In Goochland, there is an attempt to limit the stress about the assessments and evaluations and encourage the focus on engaged learning. To arrive at a place where the teachers believe and subscribe to those values takes time. To stand within teachers' fears raised by the use of assessment measures based on student achievement and ask for more takes courage. And it is all based on an environment where trust is continually built. Dr. Lane shared this:

> I don't think it's that teachers are necessarily always reluctant to make a pedagogical shift, but they're reluctant to believe that administration and leadership really means this is the way we want to go. It's not that we don't care about test scores. It's that we believe test scores will follow and it's okay to jump in and change the way you design and deliver lessons. We need to be trusted that we're going keep our word and stay the course. (personal communication, January 30, 2014)

● COMMUNITIES OF TRUST

Certainly, we are all navigating an unknown sea. Some are entering by choice and others because powerful voices say they must. Either way, leaders are the ones to whom all look for the words and the heart that will describe the destination. **Leaders are those who will inspire teachers to stay the course and will call them to be courageous risk takers, applauding the innovation that comes to life within the school.** Leaders identify the places within the regulations and the rubrics and the rules where flexibility resides and seize upon them to open doors. Compliance is essential, but creativity is life giving. Goochland leaders demonstrate this.

Human beings gravitate toward authenticity. This may reveal why really good teachers become such strong leaders. Many of them have mastered this ability to reside in and act from the heart. That is the pool from which leaders have come. **Leaders benefit from remembering themselves in that previous life and reconnecting to that power source. The best teachers are purveyors of hope. Educational leaders must be that also. It is needed now more than ever if we are to realize the full potential of a STEM shift.**

There are many routes to this end. Creating an environment in which a sense of community, risk taking, coaching, growing, and learning take place as a matter of the operational plan is tough work. The leadership team charged with creating and maintaining such an environment can easily be pulled toward the to-do list for the implementation. It is a natural pull for several reasons. Accomplishing tasks and crossing them off a list is immediately rewarding. Putting the strategic plan up on the website, preparing the fliers for distribution, having open and honest meetings, encouraging risk taking, and keeping your promise to accept failures as bumps in the road are all fundamental. But if this environment wasn't in the school or system previously, there is a good chance that building and nurturing this environment may not be in the value set or the skill set of the leaders and teachers. This calls for more than traditional professional development.

> Pressures of immediate tasks and the bottom line often crowd out personal needs that people bring into the workplace. Every organization is a family, whether caring or dysfunctional. Caring begins with knowing about others—it requires listening, understanding, and accepting. It progresses through a deepening sense of appreciation, respect, and ultimately, love. Love is a willingness to reach out and open one's heart. An open heart is vulnerable. Accepting vulnerability allows us to drop our masks, meet heart to heart, and be present for one another. We experience a sense of unity and delight in voluntary, human exchanges that mold "the soul of community" (Bolman & Deal, 2008, p. 103).

It may make the reader uncomfortable to talk about openheartedness in the process of gathering information about a STEM shift, but, wholeheartedly, we assert that this book has caused us to encounter many of those STEM-shift leaders. They have this in common. As a STEM shift begins everyone becomes vulnerable; everyone is in a beginner's place. Both can be uncomfortable experiences.

Dr. Brené Brown (2012), author of *Dare Greatly: How the Courage to Be Vulnerable Transforms the Way We Live, Love, Parent and Lead*, is the current expert on vulnerability. She suggests that education needs "disruptive engagement . . . to reignite creativity, innovation and learning." Further, she proposes that "leaders have to re-humanize education and work" in order to create an environment in which teachers are willing to take great risks and are asking students to do the same (p. 187). It is with open hearts we truly understand and communicate understanding of what it feels like to try something new and fail, in public. This is not the environment most of us have mastered. When failure can become the proudly received invitation

> When failure can become the proudly received invitation to try again, to get better, to overcome, to be fully who we are, then we have accomplished more than a STEM shift; we have created environments where all children can learn and where all adults are doing the work to which they have been deeply called.

to try again, to get better, to overcome, to be fully who we are, then we have accomplished more than a STEM shift; we have created environments where all children can learn and where all adults are doing the work to which they have been deeply called.

> A safe environment for the learner (and for the teacher) is an environment in which error is welcomed and fostered—because we learn so much from errors and from the feedback that then accrues from going in the wrong direction or not going sufficiently fluently in the right direction. (Hattie, 2012, p. 19)

Metro Nashville Public Schools

As with all journeys, there are many routes to this end. One such journey was taken in the Metro Nashville Schools. Vicki Metzgar, then STEM Director of Metro Nashville Public Schools and now Director of Middle Tennessee STEM Innovation Hub, developed a design for building communities of trust within the newly created STEM Magnet Schools of the MNPS System. The three schools—Hattie Cotton STEM Elementary, Bailey Middle School, and Stratford STEM Magnet High School—were designated to develop a collaborative K–12 relationship. This allowed students moving progressively from one school to the next to have vertically planned learning experiences with increasing complexity. Relationships had to be developed among teachers and instructional design leaders within and across the schools. The plan was based on the research of Bryk and Schneider (2002), Tschannen-Moran (2004), and the writings of Parker Palmer (2007) in *The Courage to Teach*. The Building Communities of Trust workshop series created the environment in which teachers and leaders shared a safe space for professional growth, collaboration, and reflection, leading to the success of students, both academically and socially.

The plan was implemented over two years with facilitators prepared by the Center for Courage & Renewal. The Center was founded by Palmer and others and utilizes the Circle of Trust® approach. Invitational workshops were offered four times each year. A core group of teachers and teacher leaders participated throughout the series; others participated intermittently, attending only one or two sessions each year. The workshops were conducted with the Center for Courage & Renewal's touchstones as guidelines for establishing a community of trust.

Participants reported discovering what a community of trust was and how it was different from previous work environments

Center for Courage & Renewal Circle of Trust® Touchstones

Link also available at http://bit.ly/ TheSTEMShift

and supportive of stronger relationships and better communication. One early childhood teacher posted the touchstones on her classroom wall and introduced them as the guidelines to establish her classroom environment. Others found value in learning the skill of asking open, honest questions. As one elementary teacher wrote,

> I believe that a direct result of me attending this experience has allowed my own classroom to become a community of trust. My students created relationships with one another that fostered respect and accountability. My relationship with my colleagues has been enhanced in a number of ways, ranging from dedication to our students, to loyalty to one another. We have made a commitment to hold ourselves and one another accountable to create meaningful relationships. (personal communication, February 2013)

Since the STEM environments were new to most of these teachers, it was important that one observed that "through all the storms that are the operation of schools, this team is the 'rock,' the constant, and the most trustworthy thing here."

Absent an intentional experience like the one that took place in Metro Nashville, STEM-shift leaders have to figure out how to create an environment in which teachers and fellow leaders feel safe enough to take an ultimate risk, trusting each other. STEM shifts are deep organizational shifts, impact every child, and change teacher practices. Leading through this requires leaders express and act with compassion, empathy, and support in order for risk taking and open sharing to take place. Changing practice requires courage on the part of the practitioner and the leader. The successful shift in teaching and learning grows over time. It is a dynamic shift that requires open minds, open hearts, and open wills. Without doubt, it is hard work that requires a supportive, encouraging, safe environment and it is rewarding.

● TECHNOLOGY AND TRUST

The introduction of new technologies further exacerbates the teacher and leader vulnerability and the need for risk taking. An example of a place where trust can become fragile is found in a story about a teacher who volunteered to transform his face-to-face class into a blended-learning opportunity in which his students completed part of the course outside of the school day. With the full support of his principal, the course was designed and implemented. Unlike with traditional courses, when online learning is designed, not only does it take a great deal of time to design the lessons, it takes additional time to create the lessons as digital experiences for the students. Once the course is developed, there are ongoing communications between the student and teacher within the blended environment, and a lot of feedback and grading in a time environment without boundaries of walls or bells.

Because this was an online course, not delivered in a classroom, the principal does not see the teacher interacting each day with students, or sitting in the faculty room grading the work produced by the students. Online or blended learning makes much of the work of the teacher invisible to the untrained or unknowing. Whether absent from school or present, the teacher of the blended-learning course in this example is able to review student progress and communicate with his students, unseen.

Traditionally, communication between teacher and student is visible. Unless a principal or supervisor knows how to be an online observer of that course, the observer cannot develop an understanding or truly evaluate the teacher's work. In this case, the principal's lack of understanding of online work resulted in an unintended consequence—the perception of the principal, and those untrained in this teaching and learning medium, was that the online work had less value than the face-to-face work done by the teacher.

Without the leader's understanding of these new vehicles for teaching and learning and experience using them themselves, teachers can be more isolated, devalued, and shut down. That is no one's goal. As Principal Michael Steele suggests,

> My recommendation for leaders is for them to really think outside of their comfort zone and learn how to develop a passion to serve other people. Even the people that work for you—serve them; don't tell them what to do without sitting down and doing it yourself.

REFERENCES

Bolman, L. G., & Deal, T. E. (2008). *Reframing organizations: Artistry, choice, and leadership* (4th ed.). San Francisco, CA: Jossey-Bass.

Brown, B. (2012). *Dare greatly: How the courage to be vulnerable transforms the way we live, love, parent and lead.* New York, NY: Gotham Books.

Bryk, A. S., & Schneider, B. (2002). *Trust in schools: A core resource for improvement.* New York, NY: Russell Sage Foundation.

Covey, S. M. R. (with Merrill, R. R.). (2006). *The SPEED of trust: The one thing that changes everything.* New York, NY: Free Press.

Hattie, J. (2012). *Visible learning for teachers: Maximizing impact on learning*. New York, NY: Routledge.

Palmer, P. J. (2007). *The courage to teach: Exploring the inner landscape for a teacher's life*. San Francisco, CA: Jossey-Bass.

Tschannen-Moran, M. (2004). *Trust matters: Leadership for successful schools*. San Francisco, CA: Jossey-Bass.

Tschannen-Moran, M., Parish, J., & DiPaola, M. F. (2006). School climate and state standards: How interpersonal relationships influence student achievement. *Journal of School Leadership, 16*, 386–415.

Wheatley, M. J., & Kellner-Rogers, M. (1996). *A simpler way*. San Francisco, CA: Berrett-Koehler.

RESOURCES

 Access live links at **http://bit.ly/TheSTEMShift.**

Center for Courage & Renewal. *Circle of Trust Touchstones*: http://www.couragere newal.org/wpccr/wp-content/uploads/touchstones-poster.pdf **(QR code on page 136)**

13 A Call to Action

If you want to build a ship, don't drum up people to collect wood and don't assign them tasks and work, but rather teach them to long for the endless immensity of the sea.

—Antoine de Saint-Exupéry

Leading a systemic shift is far more complex than instituting a new program. The system's mission, its values and culture, its organization, and the context of law and regulation that defines it, are key factors that inform the choices and decisions about how a specific shift will be planned and implemented. We understand the intricacies of our systems and ourselves, and thus, our path. The STEM shift, as described throughout this book, touches all levels and aspects of teaching and learning—subjects taught, methods used, schedules, relationships, communication, affiliations, support, and the very nature of how business is done in our schools. This is all done in the service of leading education systems into and beyond the punctuated equilibrium that is disrupting our old patterns and allowing for 21st century schools to emerge.

This shift calls on the leaders to keep their sights on the long view without losing agility on the political, fiscal, and educational aspects of the day-to-day job. Pausing for moments each day to see the big picture and restore the dreamer within can become sustaining daily practice. After those moments, we encourage leaders to examine to chart below and commit to actions each day that contribute to one of these essential contributions leaders must make for a successful to shift to emerge and grow. Leader decisions and actions in these five areas support followers to do the same. Ultimately, an environment is created in which students, the ones to whom our lives are professionally dedicated, become engaged learners, who are also risk takers and relationship builders, STEM skilled and full of heart.

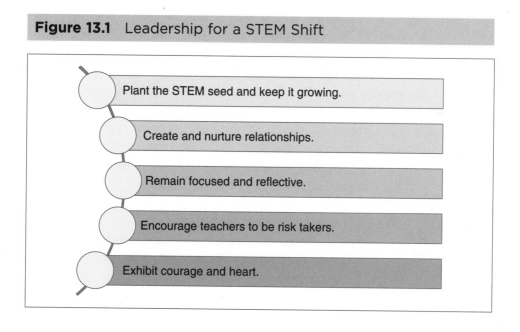

Figure 13.1 Leadership for a STEM Shift

Plant the STEM seed and keep it growing.

Create and nurture relationships.

Remain focused and reflective.

Encourage teachers to be risk takers.

Exhibit courage and heart.

● URGENCY

We must catch up. At this writing, we are, after all, fifteen years into the 21st century. There are big gaps in our student achievement results and resources. Many observers are demanding alternative and parallel systems because they see public education as too entrenched, too privileged, and too insulated. The good news is that STEM has arrived as a choice to embrace. Others, in schools, corporations, and universities across the country, already have.

Educators have spent years tinkering with a system and calling it change. We have been working primarily on the surface and the edges of our system. **STEM calls us to enter the territory beneath the surface and question the structural and organizational assumptions we have made.** It invites us to unify what has been artificially separated within schools, separated by labels, by disciplines, by certification areas, by identities and expertise. The separations are further defined externally by school and district boundaries. All around us the world presents examples of this question: Are we separating more or finding ways to connect? Technology has made us so much more connected and opened all kinds of doors. How rapidly its capacity changes our work will be an ongoing question. But the time has come for us to take the dive into the fundamental questions.

We regret the term *STEM* is perceived to be limiting and excluding. However, a reason to hold on to the acronym is the public's view of the name, no matter how misunderstood, and its value for the 21st century schools we need to design. Partnerships, money, scholarships, and professional development and facilities follow its adoption.

STEM can appear as a program added within an existing system or as a club on its periphery. Wherever it sprouts, STEM is the vehicle that can shift the very foundation and structure upon which our schools have been built. It is important we respect the complexity and dynamic nature of the challenges this shift presents. It incorporates the social sciences and the arts as a systemwide change in every aspect of teaching and learning and the nature of how we do business.

> Wherever it sprouts, STEM is the vehicle that can shift the very foundation and structure upon which our schools have been built.

There are schools already on the path to the shift. Some don't yet recognize the potential they have invited inside. Others do; they have supported their pioneers and created pilot programs intentionally. We can learn from them. There are extraordinary efforts being made by district leaders along with teachers and communities to design a comprehensive new vision for schools. Impressive professional development is being offered, and partnerships are growing. Teachers have taken it upon themselves to sweep away old classes and create their own small-scale STEM shift. In schools like those in Goochland County, Virginia; Nashville, Tennessee; East Syracuse, New York; and Salt Lake City, Utah, we've seen examples of STEM-system shifts that lead the way. From their work, we are encouraged.

● THE STEM SHIFTERS

Along the investigative journey to do research for this book, we discovered school districts in which boards of education, superintendents, assistant superintendents, directors, and teachers were synchronized in the shift process. They had a vision and a strategic plan to follow as their shift unfolded. They understood the power of the shift and generated the energy of its momentum with learning opportunities, coaching, and a risk-accepting environment in which to try and succeed or fail and try again. They partnered with corporations, STEM hubs, universities, and others.

From Goochland County Schools in Virginia, Superintendent James Lane, Assistant Superintendent for Instruction Steven Geyer, Director of Career and Technical Education Bruce Watson, Supervisor of Instructional Technology John Hendron, Principal Sandra Crowder, board member Elizabeth Hardy, and fourth-grade teacher Beth Ferguson all shared their collaborative success in developing and following their strategic plan. They designed and follow a G21 plan that provides the framework for how they will develop 21st century skills for their students. Together, they know how to acknowledge efforts and celebrate successes. Ferguson's excitement about the education she received from the VISTA program at George Mason University, the coaching she received as a result of that program, and the daily changes that are taking place in her classroom for her students is recognized, supported, encouraged, and celebrated by the leadership team. Her classroom environment is active and her students are engaged, problem-solving learners.

From Superintendent Donna DeSiato at East Syracuse Minoa (EMS), we learned of the long-term commitment to achieve a systemic shift to become a 21st century district. Along with strong connections and partnerships with corporations, financial institutions, and universities, an understanding of purposeful change—especially one this deep—change as a process not an action or decision has emerged. EMS has embraced the STEM shift across the entire system.

As ESM shifted the system, progressing through STEM, STEAM, and STREAM teams, the nature of what was happening in every classroom became different. They used their partnerships to maximize learning opportunities outside of the classroom and moved learning right into the lobby of the high school. Their partnership with Siemens began in 2009 as part of an energy performance contract for the district. The contract morphed into a relationship. Together they applied for and received a New York State Energy Research and Development Authority (NYSERDA) grant. That grant allowed them to provide a kiosk in the lobby of the high school where students learn about energy use in the district, including solar energy. Students are active learners searching the source of energy being used at any point in time anywhere in the district by using a touch screen. A portable device available in the elementary and the middle schools offers similar data. "This began a dialogue and relationship that led to what we refer to as a "partnership for learning." Since that time Siemens has been a partner in our development of STEM/STEAM/STREAM models for learning and the shift to trans-disciplinary, project-based learning," says Dr. DeSiato. ESM maximizes their partnerships as resources for professional development for their teachers and as teaching partners, both inside and outside the classroom. And ultimately,

Teachers are at the center of the shift in the design of learning models for 21st century learners and outcomes.

Dr. DeSiato credits her "teachers who are at the center of the shift in the design of learning models for 21st century learners and outcomes."

There are examples of schools in Nashville. Hattie Cotton STEM Magnet Elementary School, where 92 percent of students receive free and reduced lunch and 82 percent are minority students, was able to increase reading and math scores after the shift to STEM. At Stratford STEM Magnet High School, with 92 percent of students receiving free and reduced lunch and 77 percent are minority students, students who were previously disengaged or passive learners, with little or no confidence in their academic abilities, stepped up and some became award and scholarship winners. There is evidence at DuPont Hadley Middle School, also in Nashville, that the shift to problem- and project-based learning, caused scientific inquiry to become a sound foundation for practice. As such, their Cell-a-Bration is a testament to what can happen when the right training, support, encouragement, and time are in place for the teachers. This team of teachers who trained, planned, and taught together, facilitated the environment in which the students learned about cells (science), multiplying fractions (math), conducting research (reading), and persuading adults to support their fundraising efforts (writing). Students developed their public speaking skills, as they were required to speak in front of a live audience (presentation). Students' interest in the learning was reinforced by the authentic project selected and designed by teachers. It related to their concern for their classmate who was a cancer survivor. Their motivation was to raise awareness and funding so that a cure could be found. "They will never forget this experience and all they have learned" (teachers, Pam Newman and Caroline Neuffer).

Eric Sheninger, who during the research for this book was principal of New Milford High School Academy, described the academy program and the way it helps students find their passion, their path, whether for college or career. The goal is for students to explore and pursue their passion before leaving high school.

Tyler Howe, principal of Neil Armstrong Academy in Utah, reinforced the importance of a districtwide vision for the planned STEM shift. They began by building their STEM elementary school; middle school will be next, then the high school. He emphasized the importance of working with the faculty to come to consensus about what it means to make a STEM shift and the care it takes to stay the course. These are only a few of the schools and districts across the country from which to learn.

● LOCAL SEEDLINGS

The seeds of the STEM shift already exist in most schools. Classrooms and teachers are successfully embracing the fundamentals of project- and problem-based interdisciplinary instruction. They are embedding them in their work with students, sometimes without notice.

Philip Pietrangelo, a chemistry teacher working in the Diman Regional Vocational Technical High School in Fall River, Massachusetts, worked with a colleague to open their doors and lower the walls between their classrooms and their subjects. Their belief in the power of changing learning experiences for their students fueled their dedication to make it happen, even if it wouldn't fit into the existing schedule. Their club/class hybrid for credit resulted in a program that allowed their students to learn chemistry in the context of the careers they were hoping to pursue. It also resulted in a noticeable difference in their level of engagement when attending their traditional chemistry classes. The authentic connections had been made through their Food Science club. Philip no longer works at Diman Regional; he is now at Boston Green, where he continues to look for ways to replicate his approach to the authentic integration of science and the other subjects his students study.

George Mayo (2014), a high school film and English teacher in the Montgomery County Public Schools, Maryland, took it upon himself to combine literature with architecture. On his own, he connected his students with professors and college students of literature and architecture and helped his tenth-grade class think in new and creative ways about literary thinking. What if he had a 3D printer? If that district chooses to make the STEM shift, Mayo would most naturally become one on the design team as the district begins its shift.

> The seeds of the STEM shift already exist in most schools.

● COLLEGES AND UNIVERSITIES

K–12 educators are not the only ones who need to make this shift. Colleges and universities must both lead and respond to the K–12 shift. This is particularly essential for schools of education, where teachers must be prepared for a shifted environment and for all undergraduate education, which STEM-prepared high school graduates will soon enter. Admission departments will be recruiting a new brand of student, one experienced in problem-solving, project-based learning and collaborative inquiry.

Students may or may not enter the university to become teachers, yet there are already scientists and professors at Vanderbilt and Columbia who find themselves working with K–12 STEM students as systems shift. The intention and the capacity of a STEM shift must, and will, reach past the twelfth grade into our colleges and universities. The STEM shift will make the concept of cradle to career an applied imperative.

An example can be seen in the work of Matteo Pericoli, the architect, illustrator, teacher and author, who taught a course at Columbia University School of the Arts called The Laboratory of Literary Architecture. It provided the impetus for George Mayo to turn his class around. Pericoli's

> course examined how certain principles of architectural design can be used to describe literary structures ... By breaking the text down into its most basic elements and analyzing the relationship of each part to the overall structure, they determined how the text could best be translated into architecture. (See Columbia University)

His work serves as a prototype as he reaches into a public school classroom. There is a wave of newly prepared students arriving on college campuses soon and they will expect these kinds of integrated courses.

A recent meta-analysis tested the hypothesis that lecturing in college classes maximizes learning and course performance (see Active learning). In all, 225 individual studies were included. The results raised questions about the value of continuing the use of traditional lecture and support for the use of active learning as the "empirically validated teaching practice in regular classrooms" (p. 1).

> In addition to providing evidence that active learning can improve undergraduate STEM education, the results reported here have important implications for future research. The studies we meta-analyzed represent the first-generation of work on undergraduate STEM education, where researchers contrasted a diverse array of active learning approaches and intensities with traditional lecturing. Given our results, it is reasonable to raise concerns about the continued use of traditional lecturing as a control in future experiments. (p. 4)

Advice to faculty in the schools of education is to require more coding, gaming, 3D printing, and blended and online learning. They might begin by becoming transsubject instructional teaching models for their students.

● INCENTIVIZING THE SHIFT

There exists a plethora of online resources for those engaged in a STEM shift. Teaching Institute Excellence in STEM can provide information and networking opportunities. Subscribing to Mind/Shift, Edutopia, STEMconnector, EdSurge, and edWeb, (see Resources) can bring new ideas, almost daily, to your computer. The National Education Association (see Resources) offers STEM resources—and hundreds more are at our fingertips with a simple Google search. Develop a social bookmarking account, such as Diigo, where you can begin to collect and share the resources you are finding relate best to your school and district. Modeling and promoting the use of technology as you enter the shift will help as information is collected and shared. Establish a Facebook and Twitter account focused on this work and follow those who are STEM thought leaders.

At the highest levels of federal government, the STEM agenda is gaining momentum. A May 27, 2014, briefing paper from the White House described the president's action plan, Educate to Innovate. Components include $35 million to launch a competitive grant program to prepare one hundred thousand "excellent STEM teachers" over the next ten years and an expansion of STEM AmeriCorps to offer STEM learning experiences with robotics, coding, and food production to eighteen thousand low-income youth in the summer of 2014. A pilot mentoring program supported by US2020 is beginning in seven cities with a goal of mobilizing one million STEM mentors annually by 2020. Corporate sponsors include Cisco, Cognizant, Raytheon, SanDisk, Chevron, and Discovery Communications. Another component is a program to give every school in America access to GIS technology to map and analyze data, give students 24/7 access to the data via smart technology and collaborate in the cloud.

> At the highest levels of federal government, the STEM agenda is gaining momentum.

STEM partners and new initiatives are popping up with a frequency that is hard to imagine. NBCUniversal's Hispanic Enterprises and Content is beginning a campaign to close the Latino achievement gap in the STEM fields. A Global STEM Alliance is designed to connect students from around the world. NASA and the Khan Academy are joining forces to make space exploration available in lessons, simulations, and games online and free. Maker Ed has a new online library for resources ready to come online, and Time Warner Cable is sponsoring after-school STEM programs. The Society for Science & the Public (SSP) offers a Broadcom

MASTERS (Mathematics, Applied Science, Technology, and Engineering for Rising Stars) program for sixth to eighth graders. The competitive program is open to those who score within the top 10 percent of SSP-supported science fairs. Schools or districts can apply for funds to conduct microgravity experiments supported by competitive funding in conjunction with the Student Spaceflights Experiments Program at the National Center for Earth and Space Science Education (NCESSE). Subaru is one of the sponsors.

The list of those who have the desire to support a STEM shift in our schools is growing exponentially. So many times over the past few decades, we have been asked or mandated to change something and have done so with varying degrees of enthusiasm. This is different. **The choice for a STEM shift is ours to make.** Regardless of the governmental and corporate support for STEM, educators are the ones who will make it real for students and, in so doing, we will change their lives and our own. **The excitement can be contagious.** Unless the vision creeps into every school and classroom, pockets of students may be served, some of them well, but it will not be a public educational systemwide shift.

As LaKisha Brinson, library and media specialist from Robert E. Lillard Elementary School in Nashville, Tennessee, states, the significance lies in the understanding that

> STEM is more than an acronym. It is a system shift. When students are exposed to STEM practices learning takes on a new meaning. STEM involves learning various content materials through the inquiry-based practices. Students are encouraged to ask and formulate questions and to be okay with not being able to answer the question right away. Twenty-first century learning skills such as critical thinking, communication, collaboration, and creativity become a part of their daily school culture and not separate instances. STEM education embraces hands-on-learning and challenges students to be facilitators of their own learning.

● BRIDGE BUILDERS

We are blessed to live in a time when technology is developing at warp speed and it is possible to place it in the hands of your youngest children. Computers, interactive whiteboards, laptops, tablets, DSL, fiber optics, satellite, wireless access points, wide area network (WAN), local area network (LAN), 3D printing—all are part of our workplaces. These modern technologies are the 21st century's pencil, encyclopedia, newspaper, library, museum, post office, telephone, movie theater and television, molds and dies. They are the 21st century sandbox in which we can create movies, music, and art, as well as learn and communicate. We can print in 3D, and so can the children.

There exists a gap between the "educational techies" and the rest of us. There are some among us who are leaning forward, on the edge of every next technology. There are those who understand the Maker Movement, attend Maker Faires, have Scratch accounts, maintain and read blogs, have participated in a MOOC, bought and are using a 3D printer, and use social media for communication. The Internet created a platform that has given birth to crowdsourcing and crowdfunding. Sharing resources and raising money are now mainstream methods for innovating, raising money, and advertising new ideas. They embrace it and have most likely funded a crowdsourced project. Who knows what's next? There are those who are interested and can use the interactive whiteboards and the handheld devices and teach online. There are those see no value in "all of this hoopla" or are standing on the side, waiting for it to slow down so they can get aboard.

It is not easy to let go of the foundation, the *terra firma* that has served the nation so well. In the STEM shift, schools and districts need leaders to be assuring and courageous. The step into the STEM frontier will help students be better prepared for this century. Schools, their leaders, teachers, and communities are being called to action.

Missing are the bridge builders. Schools need those who understand tinkering, making, coding, even 3D printing, crowdsourcing and crowdfunding, social media, and all of the extraordinary tools we now have available to use in a teaching and learning environment. Schools need to explore the relationship these vehicles have with required standards and curricula. Schools need those who can demonstrate how, not only to appreciate the capacities of these new tools, but to use them to meet the needs of students. Schools need

> We need bridge builders to step forward.

those who can encourage the use of these technologies as vehicles for learning and, "Make technology the servant, not the master" (Martinez & McGrath, 2014, p. 161). We need bridge builders to step forward.

Let us learn from the scientists Jan C. Bernauer and Randolf Pohl (2014), who tried to precisely measure the radius of the proton. In complementary experiments, their results differed greatly. The approach these scientists used to deal with their different and unexpected findings offers a model of the soft skills that must be embedded in our 21st century teaching and leading. They abandoned old beliefs and faced the evidence their findings revealed to them. They said, "We have begun to dream of more exciting possibilities." We are being called to lead schools of a new day for a century already one decade gone. To do so, we will have to face the evidence. Change pressures are upon us. We will have to abandon old beliefs about what works. Only then can we begin to dream of more exciting possibilities.

REFERENCES

Bernauer, J., & Pohl, R. (2014, January 21). The proton radius puzzle. *Scientific American, 310*(2), 32–39.

Lazlo, E. (1996). *The systems view of the world: A holistic vision for our time.* Cresskill, NJ: Hampton Press.

Martinez, M. R., & McGrath, D. (2014). *Deeper learning: How eight innovative public schools are transforming education in the 21st century.* New York, NY: The New Press.

RESOURCES

 Access live links at http://bit.ly/TheSTEMShift.

Columbia University School of the Arts. *The Laboratory of Literary Architecture: A Workshop With Matteo Pericoli* (literary architecture lab at Columbia University): http://arts.columbia.edu/laboratory-literary-architecture-workshop-matteo-pericoli

EdSurge (for science and technology news and updates): https://www.edsurge.com

Edutopia (sharing evidence- and practitioner-based learning strategies that empower you to improve K–12 education): http://www.edutopia.org

edWeb: A professional online community for educators http://home.edweb.net

Freeman, S., Eddy, S. L., McDonough, M., Smith, M. K., Okoroafor, N., Jordt, H., & Wenderoth, M. P. (2014). Active learning increases student performance in science, engineering, and mathematics. *Proceedings of the National Academy of Sciences of the United States of America, 111,* 8410–8415 (meta-analysis about effect of lecture on learning): http://iteachem.net/wp-content/uploads/2014/05/Freeman-S-Proc-Natl-Acad-Sci-USA-2014-Active-learning-increases-student-performance-in-science-engineering-and-mathematics.pdf

Mayo, G. (2014). *Creating Architectural Models of Literary Themes* (reader idea; Montgomery Blair High School English Class): http://learning.blogs.nytimes.com/2014/05/15/reader-idea-creating-architectural-models-of-literary-themes/

Mind/Shift (Launched in 2010 by KQED and NPR; explores the future of learning in all its dimensions): http://blogs.kqed.org/mindshift/

The National Education Association. *The 10 Best STEM Resources: Science, Technology, Engineering & Mathematics Resources for PreK–12:* STEM lesson resources: http://www.nea.org/tools/lessons/stem-resources.html

STEMconnector (information about STEM): https://www.stemconnector.org

Teaching Institute Excellence in STEM (TIES): http://www.tiesteach.org

Appendix

A Checklist for Evaluating the Quality of the Planning Process

A community-wide effort will develop a strategic plan. The leadership team will be prepared to answer these questions for both internal and external audiences.

- ☐ Why is STEM our choice?

- ☐ Who will benefit from it?

- ☐ What research and data have been gathered and used to inform our decisions?

- ☐ Who are the essential partners in this process? Are they potential funding agents?

- ☐ Where are we in the change process? Initial inquiry? Beginning discussions? Ready to begin the change? Along the way?

- ☐ How is the systemwide plan being developed? By whom? Are all aspects of the shift included?

- ☐ What process was used to determine priorities or phases in the shift?

- ☐ What processes have been effective in disseminating and explaining the proposed changes?

- ☐ What plans for regular communication, feedback, and adjustments are built into the process?

- ☐ How will we respond to those expressing opposition? Who will take the lead for that responsibility?

◻ How will we discover and address the individual impact to those affected by these changes?

◻ How will we attend the emotional challenges associated with the change process, such as feelings of loss and of fear from faculty members, parents, and students?

◻ Have we established a timeline in which all steps are clearly articulated, with dates for checking in, communicating out, and anticipated completion?

Index

CORWIN

A SAGE Company

Helping educators make the greatest impact

CORWIN HAS ONE MISSION: to enhance education through intentional professional learning.

We build long-term relationships with our authors, educators, clients, and associations who partner with us to develop and continuously improve the best evidence-based practices that establish and support lifelong learning.